T0319664

Contesting Crime Science

Contesting Crime Science

Our Misplaced Faith in Crime Prevention Technology

Ronald Kramer
and James C. Oleson

UNIVERSITY OF CALIFORNIA PRESS

University of California Press
Oakland, California

Library of Congress Cataloging-in-Publication Data

Names: Kramer, Ronald (Senior lecturer), author. | Oleson, James C., 1968– author.
Title: Contesting crime science : our misplaced faith in crime prevention technology / Ronald Kramer and James C. Oleson.
Description: Oakland, California : University of California Press, [2022] | Includes bibliographical references and index.
Identifiers: LCCN 2021033099 (print) | LCCN 2021033100 (ebook) | ISBN 9780520299580 (cloth) | ISBN 9780520299597 (paperback) | ISBN 9780520971264 (epub)
Subjects: LCSH: Crime prevention—Social aspects. | BISAC: SOCIAL SCIENCE / Criminology | SOCIAL SCIENCE / Technology Studies
Classification: LCC HV7431 .K73 2022 (print) | LCC HV7431 (ebook) | DDC 364.4—dc23
LC record available at https://lccn.loc.gov/2021033099
LC ebook record available at https://lccn.loc.gov/2021033100

Manufactured in the United States of America

30 29 28 27 26 25 24 23 22 21
10 9 8 7 6 5 4 3 2 1

Contents

Illustrations

Acknowledgments

This book would not have been possible without the support of many people. We are especially grateful to the wonderful people at University of California Press. Throughout this journey, Maura Roessner has been an incredibly supportive editor, maintaining faith in the project from its initial conception through to its final version. Unlike the "faith" in crime science, we hope Maura's has not been "misplaced." Madison Wetzell, Teresa Iafolla, and Jessica Moll have also been instrumental in ensuring the book's production. We loved the cover design at first glance. We are appreciative of the support and help of the many other people who have worked "behind the scenes" to ensure the manuscript would appear in its best possible form. We are also indebted to Frea Anderson, who worked as a research assistant during the book's initial stages. We would like to thank Professor Mike Davis for generously allowing us to reproduce his iconic diagram of dystopian Los Angeles, and would like to thank Jonathan Simon for his generous description of the project. Finally, we greatly appreciate the three reviewers who read earlier drafts of the manuscript and provided much constructive criticism. The book is unquestionably stronger for their contributions.

Ronald Kramer would like to thank his family, Doris, Lee, Sharon, and Glen. Mom, I do not know where I would be without all our trips to Bunnings Warehouse. Special thanks is also owed to Neera R. Jain, especially for her tenacity in keeping up with all the latest Instagram accounts dedicated to rescuing cats, bats, and wombats. I would like to thank my coauthor, James Oleson, for his patience and perseverance. We have not spoken about this yet, but as coauthor I am assuming he is willing to take responsibility for all the errors, misreadings, and unsound arguments throughout the text, even though they are—in all likelihood—my doing. Portions of this book were written while on sabbatical afforded by the University of Auckland.

James Oleson would like to thank (and blame) Ronald Kramer for asking such interesting, heterodox questions and for suggesting this book project. Researching and writing it forced me to think carefully about my assumptions and revealed new vistas within criminology. I would also like to thank the University of Auckland, which provided essential research funding for the project. My contributions to the project would not have been possible without the support of Clare Wilde, Jameson Oleson, and John, Patricia, and Jennifer Oleson.

Introduction

The figure of the master detective can be traced back at least as far as the late nineteenth century. Arthur Conan Doyle's character Sherlock Holmes could deduce the whole of an otherwise unsolvable crime by focusing his keen intellect upon a single trifle ("there is nothing so important as trifles"). Thus, from a flake of ash and the uneven wear on the heel of a boot, Holmes could conjure the criminal in vivid detail: "And the murderer? Is a tall man, left-handed, limps with the right leg, wears thick soled shooting boots and a grey cloak, smokes Indian cigars, uses a cigar holder, and carries a blunt penknife in his pocket. There are several other indications, but these may be enough to aid us in our search."[1]

Like magic. Sherlock Holmes introduced the use of tobacco ashes, fingerprints, serology, and printed type just as they were being accepted as legitimate techniques of forensic investigation. It is a matter of some debate whether Sherlock Holmes was more product or source of early crime science.[2] But what is beyond debate is that in the 134 years since Holmes first appeared in print, generations of investigators and super sleuths—real and imagined—have attempted to emulate his ability to solve crime using trace evidence, artifacts found at crime scenes, and reason. In fact, the admixture of such elements has become a cliché. Turn on any crime drama, and there is a good chance that you will see one or more determined detectives crack an impossible case by finding a single shred of evidence—an overlooked fingerprint, a hair, a drop

FIGURE 1. Sherlock Holmes examining a cane. *Source:* "The Hound of the Baskervilles," *The Strand* 22, no. 128 (1901): 124.

of blood, or a micro-expression—from which the whole of the case is revealed. It is such a seductive narrative that the exaggerated power of forensic science has become enshrined in the public imagination. Fingerprints, DNA, lie detectors, offender profiling, and magnetic resonance imaging scans are all new forms of magic, unabashedly invested with the power to resolve even impossible-to-solve crimes. But like other forms of magic, crime science depends upon a suspension of disbelief.

CRIME SCIENCE

We understand crime science to be a broad, encompassing, discursive formation. That is, we understand it as a framework constructed from statements, categories, and relationships between concepts, all of which produce criminalized behavior and render it comprehensible in particular ways. This discursive formation of crime science is not monolithic. Rather, it is "rhizomatic," functioning as a conceptual grid that underpins a variety of seemingly unrelated offshoots: biological evidence, actuarial assessment, security technologies, and so on. Thus, crime science operates like an archipelago, in which a single formation connects a mountain range, creating what from above the water line appears to be a chain of separate islands. Most scholars focus on the individual islands (manifestations) of crime science, and there are excellent reasons for engaging in such heterogeneous critiques. But it is also crucial to examine the unifying characteristics that interconnect these islands and to identify the central ideological tenets of crime science.

The contours of crime science remain inexact, which is unsurprising since the term "lacks a standard definition."[3] Nevertheless, it is possible to eke out a general meaning. Ronald Clarke distinguishes crime science (which is narrowly focused on crime reduction) from mainstream criminology (which is interested in understanding and explaining crime). In his view, criminology strives to understand the *why* of crime (the "criminal"), while crime science focuses on the *how* of crime (the enacting of crime). Whereas criminology attempts to prevent delinquency and reform offenders, crime science seeks to inhibit crime. Criminology concerns itself with long-term social reform, but crime science is interested in immediate crime reduction. Criminology views crime as pathological and opportunity as a secondary consideration (against preeminent distant causes); crime science views crime as normal and opportunity as the central cause. Clarke suggests that the constituent fields of criminology are sociology, psychiatry, law, biology, and genetics; the constituent fields of crime science are environmental criminology, geography, ecology, behavioral psychology, economics, architecture, town planning, computer science, engineering, and design. Criminology draws upon the talents of stakeholders in the criminal justice system, social policy makers, social workers, university teachers, and intellectuals. On the other hand, crime science builds upon the talents of police and the security industry, crime policy makers, architects, town planners, city managers, crime and intelligence analysts, and design, business, and industry.

Several aspects of Clarke's description of crime science are noteworthy: its similarity to environmental criminology; the focus on immediate crime reduction through deterrence and environmental design, the weight given to description rather than theory, the emphasis on opportunity, the prioritizing of victims of crime, and the partnerships with interests exogenous to the traditional criminal justice system (e.g., intelligence analysts, security industry officials, architects, product engineers, and business leaders). But defining crime science by contrasting it with mainstream criminology overlooks key elements that inform both approaches. For example, both mainstream criminology and crime science assume that "crime" possesses an ontological reality. That is, they assume that crime is an unproblematic thing that exists independently of social construction. Similarly, both criminology and crime science believe that crime can be controlled, reduced, and managed, and both believe that science and technology necessarily play a key role in the control of crime. Although some crime scientists posit that crime should be controlled through biological interventions (e.g., Lombroso), while others suggest it should be controlled through a rational, calibrated system of reward and punishment (e.g., Beccaria), there is near-unanimous agreement that crime can be, and should be, controlled. Therefore, instead of understanding crime science as a field that resides in opposition to mainstream criminology, we see it as subsuming most strains of criminology. Indeed, crime science is not merely one discourse among many. Rather, it occupies a place of prominence in our social, political, economic, and epistemic arrangements and has become the dominant narrative in matters of crime and punishment.

DISRUPTING THE DOMINANCE OF CRIME SCIENCE

Premised upon a seemingly unshakeable conviction that crime can be "solved" with the right mix of scientific knowledge and technical know-how, crime science spreads throughout social space. We become bound by a cultural environment saturated by the belief that society could be rid of crime once and for all if only we had better practical strategies and technologies. In this, technology is imbued with magical properties, fetishized. A seemingly endless array of technologies and devices is developed and introduced, promising to keep criminalized behavior at bay.[4]

Despite its (epistemic, cultural, political, social, and economic) authority, crime science rarely appears before us as a neatly encapsulated entity. Instead, it is broken up and scattered in various ways, akin to scriptural writings not yet compiled into a single, canonical text. Nevertheless, it can be traced via its discrete domains or divisions, such as "biological," "actuarial," and, among others, "security" science. It is precisely through a critical examination of these fragments—that is, particular branches and criminological writings—that we interrogate crime science as a master discourse for framing criminalized behavior and for legitimating specific approaches concerning the control of crime.

By no means are we the first to call crime science into question. It has certainly been criticized and problematized by others. Much of the extant critique, however, tends to be piecemeal and focused on specific technological developments within crime science. Therefore, there is a tendency to ask and answer particular types of questions. Much of the existing scholarship adopts empirical methods to ask whether a technological strategy (or strategies) significantly reduces crime. Another common approach entails considering whether specific developments within crime science can be squared with legal-ethical principles.

The former mode of critique is most evident in sociological narratives, which often construe technologies as destined to failure because they leave intact underlying structural relations that are thought to engender criminal motivations. Such arguments have been leveled against closed-circuit television (CCTV), target hardening, and, among other technologies, risk assessment instruments that purport to predict future behavior.[5] The latter mode of critical analysis seems especially prevalent among legal scholars. For example, those interested in post-9/11 forms of surveillance, the rise of security as a private industry, broken windows policing, biometric identification techniques, and the like tend to focus on the effects on civil liberties and constitutionally protected rights, thereby calling into question the state's legitimacy.[6]

These critical approaches to crime science are certainly important, and we draw from them throughout this book. Both modes are, however, limited in at least two ways. First, they tend to treat the relationship between a particular technology and its effects as the fundamental problem posed by crime science. In empirical sociological critiques, there is a tendency to assume that the desirability of any given technology hinges on the extent to which it reduces some type of criminal offending. In legal-ethical analyses, the acceptability of a technological strategy increases the more it can be brought into harmony with juridical principles. Common to both styles of critique is the tendency to elide the possibility that crime science is *the problem* with which we ought to be concerned. In other words, very rarely does one encounter a *general critique of crime science*. Instead, there is a propensity to examine some particular piece of the puzzle. This is understandable, of course, given the need to critically interrogate the emergence of new frameworks and practices concerning crime control. Nevertheless, we believe that crime science is a distinct object, one that warrants analysis in its own right.

Positioning crime science as a master discourse, this book argues that it is replete with "performative effects."[7] Writ large, two broad effects are discernible. First, crime science inscribes criminality in particular ways. Most notably, it locates criminality in or on the body, especially those bodies that are compelled to live life at the margins. As a corollary to this, one finds a refusal to inscribe criminality on institutions or sociocultural structures. Second, crime science is key to understanding the very conceptualizations of "crime" with which we operate, and thus the types of crime control practices that will come to be regarded as "normal," "natural," or "self-evident." To be sure, particular manifestations of crime science have discrete performative effects, but such effects are comprehensible as variations on these two themes.

To critically interrogate crime science, we utilize three broad strategies. First, we focus on the imminent contradictions and inconsistences that riddle crime science. There are many contradictions, but one of the most profound and recurrent concerns how the technologies that crime science advocates typically presuppose the inevitability of what are regarded as self-evident manifestations of crime or criminal behavior. Often the technologies of crime science necessitate—and can only be further developed by—the persistence of the very crimes they promise to address. Biometrics such as fingerprinting and DNA profiling, for example, evince the first of these tendencies. Both require the construction of databases, tools to gather samples from crime scenes, matching techniques, and so on. But rather than eradicate crime, they must wait for it to transpire in order to be of practical utility. The prediction instruments associated with actuarial science aptly demonstrate the second tendency. Improving the accuracy of prediction requires more "data points," such as recorded instances of crime or recidivism among those paroled from prison. This, however, is exactly what the instruments are supposed to prevent.

Second, and notwithstanding frequent assertions that it is value-neutral, grounded in the prescriptions of "objective science," and so on, we explore how crime science is ultimately complicit with power asymmetries. This complicity is often embedded in its imminent contradictions, but it can also be discerned in the crime control practices it endorses. Its primary objective is the regulation of what are perceived to be undesirable behaviors and practices in light of contemporary sociopolitical relations. Crime science is enlisted to serve the interests of those with wealth and power; to that end, it assists in the violent subordination of marginalized groups. This is akin to the engineer who hopes to develop a device that can trap carbon emissions but refuses to question the logic of manufacturing cars en masse and the normalization of automobile dependency. In this, crime science fails to entertain alternative narratives, especially those in which crime is understood as a product of defective social relations or a signifier that actively constructs our sense of reality.[8] On some occasions, it goes out of its way to disqualify such alternatives.[9] These latter narratives—sociological and constructivist—are not easily reconciled, but both operate from political standpoints that suggest different practices for the controlling of "crime."

Third, we focus on the many and varied performative effects with which crime science is awash. Crime science inflects important concepts with new meanings or substantively reworks them. Sometimes, fundamental concepts and principles of justice are more or less made redundant. Ethically, if not legally, questionable practices are instantiated. To provide some brief illustrations of these tendencies, DNA technologies impact understandings of "beyond reasonable doubt" in legal contexts and the meanings attributable to other forms of evidence. In our view, DNA fractures the notion of beyond reasonable doubt, thereby skewing the public's sense of when and where relative certainty in legal judgments is necessary. Furthermore, biotechnologies can sap the power from important legal concepts. DNA dragnets, for instance, are antithetical to "probable cause" and "freedom from self-incrimination."[10] Risk prediction, especially when it is used in the context of informing the length of prison sentences, cannot but mean that an individual's freedom is unjustifiably sacrificed. Were it not the state sending the individual to prison, such practices would readily be framed as "false imprisonment," or perhaps "kidnapping." On a related note, crime science often relocates behaviors under the rubrics of "crime" or "deviance." In this sense, rather than eradicate crime, it actively creates new classes of criminal behavior. The Transportation Security Administration's (TSA's) criminalizing of jokes about bombs and weapons at airport security checkpoints evinces this possibility.[11] And, as we explore in chapter 2, some developments in DNA technologies construct as deviant those who insist upon the importance of civil liberties and rights.

Despite such limits, crime science remains popular and will be very difficult to dislodge. Much of its appeal resides in its ability to foster the preservation of contemporary power arrangements—arrangements that much of the public is ideologically invested in. Crime science panders to this. It also receives widespread support because it is able to draw from cultural narratives in which "science" figures as objective, efficient, and capable of mastering the natural world. On a more mundane level, it also occupies a prominent place in our epistemic universe because it can be exploited for economic profit. Somewhat ironically, the public may be heavily invested in crime science, but it is not well served by it. People are effectively asked to invest through taxation in a series of strategies and technologies that cannot deliver what they promise. In terms of society's most marginalized members, crime science tends to reinforce, if not cement, their permanent exclusion from social life.

THE STRUCTURE OF THIS BOOK

The discourse of crime science is explored through five substantive chapters. Chapter 1 serves two important functions. It begins by charting some of the major positions in the debate over discourse and articulates the sense in which we deploy it to interrogate crime science. The focus is on radical constructivism, successor sciences, situated knowledges, and constitutive criminology.[12] All of these frameworks center discourse as an analytic category but push it in different directions. We suggest the possibility of a "suspended constructivism," which would operate from the zones where the aforementioned frameworks overlap. In short, we are wary of the flight into "relativism" or efforts to return to "objectivity," preferring to work from the premise that discourse and practice are always mutually embedded.

The second part of chapter 1 describes the fundamental tenets of crime science. We map the broad parameters of crime science through an examination of Cesare Beccaria and Cesare Lombroso, two figures who are regarded as central to the emergence of criminological thought. Beccaria and Lombroso are often pitted against one another. They theorize crime in different ways, and their views on the purposes of punishment are, to say the least, disparate.[13] Given such divergences, they have imparted to criminological thought two distinct theoretical legacies. But despite their differences, Beccaria and Lombroso have much in common. These shared views can be regarded as the basic contours of crime science, and by excavating them from what are typically regarded as competing accounts, we seek to show that although there are disagreements within the world of crime science, such oppositions can coexist within a single, overarching logic.

To be more specific, we employ Beccaria and Lombroso to show that the logic of crime science is constituted by three core tenets. First, crime is accorded ontological status and, from here, can be understood to follow from some defect in human nature. For Beccaria, it is the human capacity for reason; for Lombroso, it is something amiss in human biological makeup. Second, both figures fetishize technology and construe it as holding the solution to crime. We use *technology* in quite a broad sense to refer to the practical application of "scientific" knowledge. Beccaria suggests that a comprehensive body of laws—applied in objective or "mechanical" ways—will resolve crime problems; Lombroso posits that a scientific taxonomy of criminal types can be discerned and subsequently used to control individual offenders. Finally, both figures accept the state as a legitimate wielder of the power to punish. They thereby fail to interrogate the power asymmetries within which behaviors are constructed as criminal and regulated.

Chapter 2 begins our examination of what could be described as more particular, discrete offshoots of crime science. It examines bio-forensics and the renaissance of biological criminology.[14] Concerning the former, two specific technologies occupy center stage: fingerprinting and DNA profiling. The fingerprint is undoubtedly one of the most fetishized objects within crime science. It has successfully been constructed as unique to every individual and infallible as crime scene evidence; comparing fingerprints on file with those retrieved from a crime scene is generally regarded as a scientifically objective, and thus unerring, procedure.[15] DNA profiling has emerged in recent times as a "superior" biometric relative to fingerprints. Much like the fingerprint, it has quickly come to be surrounded by an aura of infallibility.[16]

But fingerprint and DNA technologies are far from "magic bullets" that will solve crime problems. As we suggest, they are plagued by contradictions that undercut the promises they make. Rather than eradicate the behaviors that they treat as self-evidently criminal, bio-forensics must "lie in wait" for them, their practical utility generally restricted to retroactive interventions.[17] Bio-forensics is also inconsistent in its assumptions about human reason. Inasmuch as bio-evidence is employed prospectively to deter criminal activity, it assumes a rational actor. These rational actors, however, are very possibly also rational enough to use countermeasures to defeat bio-forensics. And when bio-forensics operates retroactively, it assumes an irrational actor: individuals who engage in crime despite a bio-surveillance infrastructure. Of course, assuming an irrational actor undercuts the notion that bio-forensics possesses deterrent value.

Concerning their relation to power asymmetries, the databases associated with bio-evidence are inevitably developed within specific socio-historical relations. As such, the profiles that they contain are skewed by class and race dynamics, which not only limits their practical utility but, more disconcertingly, intensifies control over marginalized populations. The nexus between bio-forensics and power asymmetries is further revealed when one considers how the former intersects with the legal notion of "beyond reasonable doubt." Most notably, DNA evidence implies that particular standards for establishing guilt can govern those cases in which it has some utility, while other standards—typically well below the thresholds associated with "scientific truth"—can be applied elsewhere. Rather than "scientize" criminal justice, as some scholars have argued, the effect of this is to treat as tolerable erroneous findings of guilt in cases deemed "less serious" because DNA evidence has no bearing on them.[18]

In the latter part of chapter 2, the focus shifts to biological criminology, especially its most recent renaissance under the name "biosocial criminology." Here we chart four equivocations endemic to biosocial criminology and how these correspond to, and thus reproduce, power asymmetries. Equivocation (or profound logical inconsistency) is discernible across biosocial criminology's fundamental categories and axioms, such as "crime," "victimization," the relationship it posits between science and society, numbers and their objectivity, and "adaptation." The problem of equivocation can be illustrated briefly with the notion of "victimization." Biosocial criminology argues that victimization entails experiences that interfere with one's ability to survive and reproduce. As such, "murder, rape and theft of resources" are inherently evil crimes.[19] Yet when it must be conceded that genocide and slavery fit this definition of crime, and that they have been committed by whites against Black people, new definitions are suddenly proffered. Slavery becomes "odious" but not necessarily "criminal."[20] Black victimization is denied by consigning it to history and a pathological refusal to transcend the past. In short, white criminality is repressed to ensure the erasure of Black victimization.

The rise of actuarial science is the focus of chapter 3. Purporting to accurately predict future events on the basis of large data sets, actuarial science is increasingly used in criminal justice contexts to shape policing, sentencing, and probation decisions.[21] Those belonging to groups deemed to present greater risks of offending, or reoffending, are more likely to have their freedoms curtailed by the state when risk prediction instruments are available and utilized. We begin by exploring accounts that have tied actuarialism to incapacitation, and that emphasize how this embodies a movement away from rehabilitative ideals. For some, this is problematic insofar as incapacitation is read as punitive; for others, incapacitation signifies a diminution of punitiveness because it makes no effort to manipulate the "soul of the offender."[22] In a further strand of thought, incapacitation is seen as returning to rehabilitation, thereby forming a new, hybridized mode of punishment.[23]

Arguments along these lines are valuable, and we are sympathetic to much in them. In our view, however, they tend to miss a major corollary of actuarialism: sacrifice. We use sacrifice in the sense intimated by Giorgio Agamben, wherein it refers to the figure who is cast into a "state of exception" and can therefore be killed with impunity.[24] This logic of sacrifice, however, is repeatedly erased by proponents of prediction in various ways. It is frequently claimed, for example, that actuarial prediction is more accurate than clinical prediction, thus rendering it acceptable. Alternatively, it is claimed that if accuracy is improved, then risk prediction instruments will be more palatable. For some, the known and inevitable problem of inaccurate prediction is dismissed by asserting that some types of error—such as (falsely) imprisoning an individual who would not have offended if left in the community—"cost less" than other types of erroneous prediction, such as granting freedom to an individual who proceeds to offend.[25]

In response to such dismissals of forecasting error, we argue that even if instruments could achieve perfect accuracy—which, for a variety of reasons, they cannot—their use would remain problematic. The moral and legal dilemma is not whether one should prevent or allow future crimes to transpire, but rather a matter of determining which course of amoral—if not "criminal"—behavior is to be selected. In other words, even perfect prediction would demand that one choose between "future crime" or state-sanctioned "false imprisonment." Framed this way, the use of prediction is inherently problematic and entails harm. The problem, of course, is only exacerbated given that forecasting instruments always produce inaccurate predictions, especially in criminal justice settings, given the rarity of the events with which they are typically concerned (e.g., extreme, interpersonal violence).

Chapter 3 explores an additional problem associated with risk prediction—namely, its arbitrary, perhaps one should say political, conceptualization of data and, more specifically, how such data are to be utilized. Actuarial databases and associated risk prediction instruments are invariably used to criminalize individuals, not structures. However, the data upon which the actuarial depends could just as easily be narrativized as a revelation of the deficiencies spun from power asymmetries. In this sense, risk prediction instruments—and the knowledge of correlations embedded within them—reflect particular social arrangements, but they are used to stabilize these arrangements, not alter them. If, for example, "unemployment" is a robust predictor of "crime," we may just as well focus on unemployment, or labor markets, rather than the individual who happens to be unemployed. Refusing the first of these political options, risk instruments are directed and wielded against particular individuals. As a result, they do not liberate or promote greater freedoms, but rather work to reproduce the marginality of those who are already among society's most vulnerable.[26]

Chapter 4 interrogates the logic of security science through a focus on relatively recent anti-terrorism efforts, CCTV, and private security. What unites these strategies—and comparable practices—is their tendency to operate as a perimeter that encloses space, thereby preventing what it sees as constitutive of "crime" or "threat" from entering particular settings. One might be inclined to assume that because it seeks to control space, security science is impartial in its operation. As it turns out, this is not the case.

This is evident in the territorial logic of security science and the cartographies of crime that it engenders. Concerning the former, because security seeks to control and dominate discrete spaces, thereby keeping "threats" at bay, it need not concern itself with proximal or distant areas. As far as security is concerned, there is no need to draw a distinction between repelling from its territory behaviors deemed undesirable and "displacement" or "innovation." That is, if its territories are devoid of "crime" or "threat," it has fulfilled its mandate. What might happen in another area is a problem for some other territorial warden.[27] Concerning the latter, we suggest that security science has the capacity to disrupt conventional cartographies of crime and subsequently "reterritorialize" them in accordance with what its technologies can observe and measure. This can lead to pulling a strange mix of criminalized behaviors onto a single, horizontal plane. Airport security technologies, for instance, routinely pursue—and thus equalize to some extent—terrorists, shampoo smugglers, and those who tell bad jokes. Ostensibly installed to prevent violent interpersonal and serious property crimes, CCTV ends up being utilized to control a range of behaviors that are nonthreatening but socially undesirable.

Chapter 4 also explores the ability of security science to reveal the massive expenditures that the enclosure of space entails and the technology fetishism that engulfs crime science. The US-led "war on terror" has hemorrhaged trillions of dollars. Brown University's Cost of War project estimates that between 2001 and 2019, the US war on terror cost $5.9 trillion and—in Iraq, Afghanistan, and Pakistan alone—claimed more than 480,000 lives. Following a shock doctrine model, the security-industrial complex quickly exploited the horror of 9/11 to secure lucrative contracts in the fight against terrorism.[28] Billions were spent in the pursuit of technologies that would safeguard US interests, taxpayer money reinvested into private security interests. There are many explanations for this, such as the need for governments to be perceived as responsive and proactive in ensuring safety from terrorism. Technology fetishism—an unquestioning faith in the power of science and technology—however, stands out as a key component in explaining why the US government was conned into handing over millions in public monies to charlatans.[29]

In a final section, chapter 4 turns to what appears to be a dominant style of critique concerning security science. Critical scholarship tends to focus on how individual liberties, especially rights to privacy, are imperiled by security. From here, scholars often posit that the state could modify legislation to neutralize any overreach of security technologies or "strike a balance" between the need for security and the importance of protecting individual liberties.[30] We do not necessarily disagree with this mode of critique, but we argue that it misses a very important point and, in doing so, demonstrates how discourses on crime science overwhelmingly remain within "permissible limits." In other words, much of the critical scholarship on security evinces a respect for the cultural taboo that renders critique of fundamental power asymmetries beyond the pale of acceptability.

That the technologies and practices summoned by security science—such as warrantless wiretaps—can thrive despite the threat they pose to individual liberties and democracy makes sense if we adopt a disturbing, but not inaccurate, view of the modern state. Rather than a protector of individual liberties or guarantor of human rights, a view that is often adopted by those who demand the state do more to protect democratic freedoms, the state is a bifurcated entity. Formally, or perhaps one should say ideologically, the state promises individual freedoms, but informally, operating on its "underside" as Michel Foucault might have put it, the state sanctions a multiplicity of technologies and practices that erode whatever human rights it is prepared to formally recognize.[31] This is perhaps most obvious when one looks at recent security efforts born of the "war on terror." As Clausewitz observed, "War is the continuation of politics by other means." In the war on terror, we see a rapid movement away from a framework of "criminal jurisprudence" and toward one of "national security." Somewhat akin to actuarialism, the latter framework is coterminous with the "state of exception," in which the state can abandon individual rights and protections. State atrocities can, and do, follow.

In the fifth and final substantive chapter, we explore a variety of theories (and some of their associated practices) that can be located under the rubric of environmental crime science. Specifically, we examine "routine activities theory," "situational crime prevention," "crime prevention through environmental design," and, among others, "broken windows." What unites these approaches is their common emphasis on creating physical spaces that are imminently crime proof. We begin by exploring how environmental science contains a logic of crime control that replicates the core tenets of neoliberalism.[32] There is, for example, the recurrent claim that states are incompetent and inefficient in the provision of services. As such, the market—in this case, security firms, managers, corporations, and so on—ought to take over the crime control functions historically entrusted to the state. If this is done, benefits will "trickle down" such that all stand to gain.

Given that neoliberalism thrives on racism and classism, and because the logic of neoliberalism so thoroughly interpellates environmental crime science, it is not surprising to find such power asymmetries resurfacing in environmental science. This is particularly striking in broken windows theory, which continues to exert a powerful influence over contemporary policing practices.[33] As such, much of chapter 5 is devoted to the racist and classist underpinnings of broken windows. Moreover, and perhaps more pertinently, the chapter examines the tendency within critical scholarship to recuperate broken windows theory. Here, we discern two strategies of recuperation. On the one hand, scholars often suggest that a chasm has somehow emerged between the broken windows theory and the policing practices it inspires.[34] We find no such gap; the theory opposes constitutionally protected rights and openly acknowledges that ignoring constitutional safeguards will pave the way to racist and classist policing. If anything, the proponents of broken windows endorse a model of policing pervaded by racism and classism.[35] On the other hand, there is a tendency to suggest that broken windows is disconcerting because, if pushed too far, the state's perceived legitimacy will be threatened.[36] This view is correct in positing that a legitimacy crisis should accompany broken windows. However, it seems to miss the fact that the core problem is the relative absence of any such crisis. In our view, crisis fails to materialize precisely because broken windows theory and the policing practices it instantiates inscribe criminality on those who are marginalized on the basis of race and class.

In the remainder of the chapter, environmental crime science's claim to conceptualize the subject as a rational actor is interrogated. Rather than rational action, the enterprise ultimately operates with what we refer to as "bio-ecological" images of the subject. In this, the actors presupposed by environmental crime science can be read as close relations of those found in bio-forensics. In any case, "bio-ecological" actors are most apparent in situational crime prevention, wherein two broad kinds of action are presupposed. First, an opportunistic, weakly motivated offender is assumed. Here, the actor is posited as taking criminal advantage of environmental "loopholes." It follows that closing such loopholes—via target hardening, surveillance, "smarter" environmental design, and so on—will channel individuals away from crime, if not along normatively desired paths. The "habitual criminal" surfaces as the second actor. This figure appears as the residual that does not go away despite situational or environmental crime control measures.[37] Although the "habitual criminal" haunts situational prevention, it is discursively repressed.

These two kinds of actor are not easily reconciled with each other and, perhaps more importantly, they are difficult to reconcile with the core tenets of environmental crime science. If crime can be inhibited by removing opportunity—as environmental science never fails to posit—the residual of criminality should not surface so readily in its empirical studies, and there should be no evidence of "displacement" or "innovation." Its studies, however, invariably reveal the persistence of "crimes" that should be blocked by its measures, and its dependence on "representational epistemology" assures that it cannot discount the possibility of "displacement." That is, its quasi-experimental logic cannot control for innovations in behavior or for the relocation of "crime" to other spaces and times. We argue that these two actors underpin environmental science because they ultimately endorse punitive, draconian modes of control while leaving power asymmetries undisturbed. The former does so by suggesting that the "motivation" for crime is flimsy, or fraudulent; the latter by positing stubbornness, a "will to criminality" that ought to be broken.

The conclusion summarizes our critique of crime science and closes by advocating a constructivist criminology. This is intended as a rejoinder to the visions of criminology that are implicitly, and sometimes explicitly, promoted within crime science. Despite its variations, it could easily be said that crime science desires a criminology restricted to positivist epistemology and the pursuit of technological innovation and (micro) intervention. In biological criminology, the refrain that criminology must approximate "Science" by positing "biology" as an "independent variable" is frequently sounded.[38] Actuarial, security, and environmental sciences appear envious of meteorology, engineering, and kindred disciplines.

Constructivist logic is often criticized for its purported rejection of reality, evident in the supposed tendency to obsess over language, construing it as more significant than the capacity to observe or to stub your toe on a rock.[39] This, however, is quite a misreading. Constructivism directs us toward an examination of how discourse and practice, subject and object, are always interrelated. There is no such thing as practice unaccompanied by discourse, and no such thing as a discourse absent performative effects or practical corollaries.[40] In other words, any given reality requires its conceptual, discursive armature. Given the dominance of crime science, not to mention its interpellation by power asymmetries—that is, its profound investment in power relations, the desire to reproduce and entrench them further—interrogating its fundamental categories and concepts amounts to an indispensable, urgent task. It is precisely this type of problem to which constructivist criminology is attuned.

CHAPTER 1

A Brief Sketch of Crime Science and Its Limits

This chapter fulfills two important functions in this book. First, it outlines the epistemological stance that informs the argument and analysis that is explored throughout the text. In our view, crime science needs to be approached as a discourse. That is, it amounts to a set of statements about "crime," "criminality," and, among other categories, "punishment." These statements, however, do not simply map or correspond to an external, pure reality, but actively construct the objects of which they speak.[1]

To embrace the idea of discourse as constitutive of reality casts doubt on the view that reality exists independently of its mediation via language. Many agree that discourse is an important concept within critical thought, but the implications of the concept have generated intense debate. We chart several intellectual positions among those who recognize the critical utility of discourse and the constructivist logic it entails in order to indicate the sense in which it is deployed here.

The work of Michel Foucault is an obvious, if not inevitable, starting point for any exploration of discourse.[2] After a brief discussion of Foucault, we offer some reflections on radical constructivism, successor sciences, situated knowledges, and constitutive criminology.[3] Whereas radical constructivism pushes discourse in a relativist direction, successor sciences, situated knowledges, and constitutive criminology suggest that "partial" or "new" truths are possible. If radical constructivism strives to reject any possibility of objective accounts, the latter positions reinstate the possibility of adjudicating between "better" and "worse" accounts of reality. We posit that attempting to push discourse in either direction—that is, relativism or a "reworked objectivity"—invites as many problems as it resolves. As such, we note the possibility of a "suspended constructivism," which operates from within the zones of commonality where most constructivist approaches appear to overlap.

A schematic summary of crime science as a discourse is offered in the second part of the chapter. Descending from sea level, we trace the underside of the crime science archipelago by examining Cesare Beccaria's *On Crimes and Punishments* and Cesare Lombroso's *Criminal Man*.[4] This approach may seem odd given that Beccaria and Lombroso are often construed as providing distinct frameworks for making sense of crime and punishment. There are certainly differences between the two figures, but despite such tensions, their works contain many points of contact. It is these overlaps that effectively provide crime science with its fundamental logic, or what one may describe as the perimeter that bounds crime science.

This initial discussion of crime science focuses on four of its elements. First, it explores crime science's ontology, epistemology, and fundamental categories. Crime science is a "modernist" discourse, one that accepts the distinction between reality (ontology) and knowledge (epistemology). In accepting this distinction, crime science believes itself capable of producing knowledge that reflects reality with fidelity. Its central categories include human nature, crime, criminality, and punishment, each of which is thought to be objectively knowable and thus liable to manipulation. Second, we explore the technology fetishism that suffuses crime science. Technological strategies figure as the primary mode of intervention or manipulation that will, so the story goes, control crime and "protect" the social. Third, some of the practical effects or manifestations within the domains of law and criminal justice are summarized. Crime science may be discursive, but this does not mean it is incapable of shaping practice.[5] Finally, the ways in which crime science is complicit with state authority, and its related tendency to naturalize and protect the fundamental axes of power by which social relations are hierarchically organized, are explored.

CRIME SCIENCE AS A DISCOURSE

The purpose of this book is to interrogate contemporary crime sciences, which we regard as a particular discourse. Engaging in this kind of critique requires that the epistemological position and theoretical ideas by which it is informed are outlined in some detail. Our approach is broadly consistent with constructivist logic, in which the concept of discourse occupies center stage. In such an approach, discourse cannot be understood as a mirror that simply reflects some external world. Rather, it needs to be understood as constitutive of the world that it purports to map.

Arguably, discourse has emerged as one of the central categories for critical approaches within criminology (and a range of other social sciences), especially those associated with postmodern thought. The most obvious starting point for elaborating upon discourse is the work of Michel Foucault. In Foucault's work, a discourse is constituted by a set of statements that cohere to form the objects of which they speak. At any given time, a variety of discourses are available for speaking about the same topic. As such, the discursive realm is fragmented into competing "regimes of truth," with specific regimes securing dominance while others remain subjugated.[6]

One can draw out the distinctiveness of this approach by contrasting it with Karl Marx's notion of ideology. Marx used ideology to refer to knowledge claims that distorted the underlying realities of capitalism, thereby preserving or naturalizing capitalist social relations. Ideology, in other words, deliberately misled people into erroneous ways of perceiving the world and thus failing to interrogate its organization, inequalities, and power asymmetries. Given this, the task of the critical scholar became one of recovering "reality," articulating the "truth" of social relations, thereby freeing the working classes from their ensnarement within ideology.[7]

For Foucault, "truth" in this sense does not exist (or perhaps one should say, the notion of truth cannot be reduced to the sense in which Marx tended to use it). There is no such thing as a reality "out there" that, if revealed, will ensure freedom, liberty, and so on. Instead, we are inevitably confined to a realm of competing discourses that strive to pass for *the* truthful account of reality. Insofar as a particular discourse becomes dominant and thus functions as if it were true, it profoundly shapes practices, behaviors, subject positions, and mental frameworks. In this sense, social practices are oriented in light of dominant discourses, as "subjects" we enter into relationships with available discourses, and it is these relationships that infuse who we are and what we do.

The notion of discourse, however, has generated intense debates within critical scholarship, and it has opened up avenues that lead to disparate epistemological standpoints. Some say that the concept of discourse has not been pushed far enough, thereby robbing critical thought of any chance to marshal a profound critique of dominant knowledges. Against this, others argue that if the notion of discourse is pushed to its limits—or if it is taken "too far"—we will quickly slide into an untenable relativism. The former view is embedded within "radical constructivism," whereas the latter is discernible in "successor sciences," "situated knowledges," and "constitutive criminology."

Given that our critique is premised upon treating crime science as a discourse, it is important to trace some of the major fault lines in the debate among these competing positions. Our discussion revolves around the epistemological positions of these different approaches and how such positions translate (or do not translate) into critical practice. By exploring such issues, we can specify the epistemological space that animates the arguments and interpretations that appear throughout the following chapters. We trace the basic parameters of radical constructivism through the work of Nicolas Carrier (and some colleagues), successor sciences and situated knowledges through Sandra Harding and Donna Haraway, respectively, and constitutive criminology via Stuart Henry and Dragan Milovanoviv.

Radical Constructivism

Like successor sciences, situated knowledges, and constitutive criminology, radical constructivism rejects logical positivism. Rejecting logical positivism entails the refusal to accept that there is an external reality that, via value-free or objective observations and methods, resurfaces in scientific accounts in some "pure" or unmediated form. Rather, knowledge is always partial, stemming from an arbitrary conflation of concepts with that which has been observed. As it is inevitable that particular concepts will be selected to summarize the observable, it follows that other concepts could have been chosen. The fundamental mistake of logical positivism is to accept the fallacy that "the world as it is" can be distinguished from "the world as it is observed."[8]

For the radical constructivist, positing that there is no reality outside of discourse, that there is no such thing as an "atextual reality," entails several logical corollaries.[9] First, it follows that any given knowledge claim cannot be afforded authority or privileged over others. An "equality of positions" is inevitable. Second, speaking in "representational" terms, or assuming that the discourse one produces maps an already extant reality, is no longer viable. Third, the idea that science entails the "discovering" of laws, which enable us to develop intervening strategies or organize institutional practices in a rational manner, must be abandoned. The world cannot be understood as consisting of "variables" and causal laws that operate independently of the subject's capacity to observe. As Carrier notes, we need to "resist dreaming of a society (and of a penal apparatus) governed by God or Reason."[10] In Nietzschean parlance, such knowledges do not follow from the "will to truth," but from the "will to power."[11] To "intervene" or suggest alternative courses of action, therefore, is an effort to exercise power.

In our view, this epistemological stance is logically coherent and thus not easily dismissed as a "god trick," as Donna Haraway has claimed. Haraway used the notion of "god trick" to characterize logical positivism and radical constructivist approaches as presuming to be able to see everything while standing nowhere.[12] While the former can be said to depend upon the "god trick," it is harder to frame the latter in the same way. If it is accepted that the distinction between reality and discourse is untenable, and that different languages or "distinctions" could be used to frame an observable regularity, it does seem to follow that there is no way to determine why specific discourses should be selected—inevitably at the expense of others—to symbolically apprehend the phenomenal. Smoking marijuana can be construed as "criminal behavior," or "pleasurable activity," or a "necessity in this day and age."[13] Of course, one may disagree that all reality is inevitably textual, but once "textual reality" is accepted, the slide into relativism seems a logical corollary.

While radical constructivism might be able to defend against accusations of functioning via the "god trick," other challenges can be raised. We argue that the problem with radical constructivism is that it cannot be translated into critical practice (knowledge production/critique) without repressing the logical coherence of the epistemological position it espouses. Alternatively, it could be said that radical constructivism can only honor its epistemological coherence by restricting itself to an endless rearticulation of its epistemology. And yet radical constructivists routinely proceed to engage in critical practice, offering a variety of deconstructive readings, critiques of various concepts and frameworks, specifying the effects that follow when certain categories of thought are readily accepted, and so on.

The tension in radical constructivism is evident in several of its writings. Nicolas Carrier and Kevin Walby's "Ptolemizing Lombroso," for example, offers a critique of biosocial criminology.[14] As they point out, proponents of biosocial criminology claim that this "new" approach amounts to a revolutionizing of criminological thought, one that demands all competing paradigms concede defeat. According to biosocial criminologists, various techniques for observing the body—such as brain scans, personality tests, and genetic markers—reveal that those who engage in criminalized behavior possess distinct biological markers, which allow the criminologist to sort those with a "disposition" for "crime" from those who do not harbor any such disposition. Biosocial criminology insists that tendencies toward crime reside within the body, but that criminal behavior is a product of biological and environmental interactions. As Carrier and Walby put it, this can be formulated as the principle of "nature via nurture."[15] From this postulate, biosocial criminology is adamant that it amounts to a radically new way of understanding the causes of crime, one that rejects biological determinism yet promises to control criminal behavior by manipulating the environmental conditions that supposedly allow biological tendencies to find their expression. If only other criminological discourses, especially those indebted to sociology, would let go of their ideological worldviews—thereby recognizing the inevitable conclusions that follow from a sober analysis of the data upon which biosocial criminology rests—the discipline of criminology could finally recognize itself as a science, one replete with a compelling theory concerning the etiology of crime.[16]

Carrier and Walby take the notion of Ptolemization from Slavoj Žižek, who suggests it is a process in which "attempts are made to change or supplement [a discipline's] theses *within* the terms of its basic framework." Ptolemization is contrasted with a "true 'Copernican' revolution": only the latter amounts to introducing a radically distinct framework.[17] With Ptolemization, Carrier and Walby are able to show how biosocial criminology does not offer anything that is fundamentally new or "revolutionary": the whole enterprise is better understood as breaking with Lombroso only in its use of specific technologies to map the body.[18] Beyond techniques of measurement, however, both enterprises remain fundamentally wedded to the idea that an unreflexive biopathologizing gaze can be embraced and, moreover, utilized to proffer symptomatological accounts of "crime."[19] And, like Lombroso, this biopathologizing gaze is not without consequences: it accords the category of "crime" an ontological status, it reproduces the "ideology of immaculate data," and it posits that "race" can be regarded as a "non-social object."[20]

This critique of biosocial criminology is more or less accurate in our view. We agree that this so-called new, "revolutionary" strand of criminological thought amounts to little more than a reiteration of the fundamental planks that underpin Lombroso's work. But why bother pointing any of this out if one genuinely accepts that knowledge claims are relative and equal, that knowledges cannot be plotted along a continuum that goes from plausible to implausible, that "representational epistemology" is impossible, and so on?

Carrier and Walby obviously accept that the written texts of biosocial criminologists exist on some plane of reality, and that those texts can be understood in "representational" terms, which is evident in positing that the conceptual armature of biosocial criminology is loaded with implications, such as reforging "race" and "crime" as ontological categories.[21] The residue of "representational epistemology" is also evidenced by portraying biosocial criminology as *Ptolemizing Lombroso* rather than *revolutionary paradigm*. Moreover, Carrier and Walby's critique of biosocial criminology is an intervention of sorts. Whereas the biosocial criminologist presumably seeks to intervene in the domain of "crime control" via the body-environment relationship, the radical constructivist intervenes in the realm of thought, seeking to disrupt specific discourses that are recognized as having some capacity to instantiate practices. It seems difficult to deny that there is a "will to power" here, that some knowledges are regarded as better than others. The radical constructivist may have abandoned the dream of a society governed by Reason, but not the dream of a knowledge production governed by a particular epistemology.[22]

To be sure, we are not arguing that radical constructivism is equivalent to the epistemological positions it calls into question. It does not, however, operate without reproducing some of the tendencies that are said to define the epistemological positions it ostensibly rejects. It is worth reiterating that we do not disagree with much that is found within the radical constructivist project. For example, we are certainly of the view that one cannot maintain an absolute distinction between subject and object, epistemology and ontology, discourse and reality; we agree that "crime" cannot be accorded ontological status, and that biosocial criminology is "pseudo-revolutionary" at best. We do, however, discern a tension between the epistemology of radical constructivism and its critical practice: it would appear as though the latter is not easily reconciled with the former. The tension may be formulated by saying that its epistemology—proceeding from a refusal to distinguish epistemology from ontology—is logical, but its substantive critiques and interpretations are grounded by the political. Is it possible to find an alternative that overcomes, or perhaps escapes, this tension?

Successor Sciences, Situated Knowledges, and Constitutive Criminology

As noted, "successor sciences," "situated knowledges," and "constitutive criminology" represent counters to logical positivism, but they are also weary of the radical constructivist's tendency to slide into relativism. In rejecting the viability of disembodied, apolitical objectivism (positivism) and relativism (radical constructivism), these positions are often propelled into a quest for what could be called a reworked objectivism. In other words, there is a desire to reconceptualize objectivity such that the possibility of "truth" or rational knowledge claims is not entirely jettisoned while preserving the critical force of deconstruction. Not surprisingly, precisely how objectivity ought to be reformulated, if not squared with constructivist logic, is the subject of intense, albeit subtle, debate.

In developing the notion of "successor sciences," much of Harding's work infuses the sociology of science and technology studies with critical, feminist perspectives.[23] Sociologies of science and technology explore the ways in which society and science are co-constituted.[24] This is to reject the idea that science as an institution can somehow be exempt from scientific analysis. As Harding suggests, how incredibly ironic it is for science to assert that it cannot be subjected to the same kind of analysis and scrutiny that it constantly imposes upon the natural and social world![25]

Many important insights follow from this approach. Science is poorly understood if one accepts that it is an institution devoted to the production of "neutral" or "objective" knowledge. Instead, the work that transpires under the banner of science always plays out in a broader set of sociostructural conditions. The types of problems, theories, methods, data, and analysis strategies—in short, every element that constitutes science—will bear some relationship to the social conditions in which they emerge.[26] In her earlier writings, Harding marshaled the co-constitutive nature of society and science to demonstrate how science was a profoundly gendered institution. This was evident by considering who worked as a scientist and how labor was divided within scientific fields. However, the gendered nature of science—its masculinist biases and gaze—was also evident in its symbolic organization and material effects. That is, science ultimately served the interests of men and, somewhat obviously from a feminist standpoint, at the expense of women's interests.[27]

Correcting this masculine bias is not reducible to granting more women access to scientific fields, nor is it reducible to securing an equitable division of labor within science.[28] The problem is that science itself operates in ways that are profoundly shaped by gender. Although it might be helpful and welcome, it is not simply a matter of adding women to the enterprise. It is quite possible for women scientists to assume and reproduce a masculinist gaze. Given this, it is important to recognize that all knowledge claims are partial—they derive from occupying some particular location—and that science would be improved by opening itself up to a greater range of such partial perspectives. Adding women and others from socially marginal locations is important but will be effective only when the particular knowledges that follow from occupying subjugated positions are recognized for what they offer.[29]

For Harding, as some of her more recent work emphasizes, *objectivity* is not threatened by this kind of diversity in knowledge production. If anything, it is only through an incorporation of partial perspectives that objectivity can be accomplished. She frames the pursuit of objectivity *through diversity* as a viable way to build a science that is more democratic, one that would work to produce more equitable social relations.[30] In this, there is a return to the Vienna circle, who did not construe science as an activity divorced from political and ethical considerations. In fact, the opposite is more plausible: science will be more objective and democratic if it adopts a reflexive stance, one that considers how its knowledge production absorbs political and ethical commitments and is replete with normative implications.[31]

Haraway's notion of "situated knowledges" is not entirely antithetical to "successor science," yet it departs from it in several important respects. Grounded in a historical sensibility, or perhaps one should say an awareness of the contextual backdrop against which postmodern thought appears to have flourished, Haraway shares Harding's skepticism concerning relativism: the sudden preoccupation with discourse seems to have arisen not long after critical discourses, especially feminist arguments, emerged to counter the dominance of positivism. As Haraway puts it: "I, and others, started out wanting a strong tool for deconstructing the truth claims of hostile science by showing the radical historical specificity, and so contestability, of *every* layer of the onion of scientific and technological constructions."[32] What began as a *practical strategy* or conceptual weapon for critique, however, quickly devolved into the disempowering of critical thought. Haraway continues: "Far from ushering us into the high stakes tables of the game of contesting public truths, [radical constructivism] lays us out on the table with self-induced multiple personality disorder. . . . [W]e ended up with one more excuse for not learning any post-Newtonian physics and one more reason to drop the old feminist self-help practices of repairing our own cars. They're just texts anyway, so let the boys have them back."[33] As these passages intimate, situated knowledge regards knowledge claims as *political maneuvers*. This excludes the possibility that they can be taken as disembodied snapshots of reality. Knowledge cannot be understood to exist in a register independent of political and ethical concerns. In this context, suspicion comes to surround radical constructivism because it suddenly asserts that all knowledge is relative; no view can be accorded more weight than any other possible view. The interrogation of power asymmetries is suddenly no more or less valid than defenses of patriarchy, assurances that capitalism "delivers the goods," and so on. Insofar as it asserts relativity, radical constructivism becomes an apology for power. It is this gravitation toward a disempowering relativism, one that would allow "hostile science" to maintain its dominance, that situated knowledge sets out to resist.

The difficult and paradoxical nature of situated knowledge is recognized by its proponents. It entails nothing short of simultaneously acknowledging the constructed nature of reality, maintaining reflexive awareness of how one's own position shapes the kinds of interpretations one is inclined to offer, and accepting that knowledge production—at the end of the day—must nevertheless strive for an account that retains fidelity to the "real" world.[34] In Haraway's view, a reworked notion of *vision*—one that is premised upon accepting "the embodied nature of all vision"—can go some way toward reconciling these contradictory demands.[35]

The rupture with "successor science" emerges here. For Haraway, this embodied vision posits a subject that may certainly be subjugated, but this in itself does not automatically guarantee "better" knowledge production. Rather, Haraway's observer must rely on technologies of vision (e.g., human eyes, microscopes, etc.) and is always fragmented or "split," and thus multiply situated. In other words, if successor science posits that those occupying subjugated social locations will "see better"—the proletariat is positioned such that they should "see through" the ideologues of capitalism; the oppressive structures of patriarchy should be more visible to women as an oppressed group; and those who are marginalized on the basis of race are positioned in a way that more readily allows for "race" to be recognized as a constructed category, one that works to normalize power relations—situated knowledges asserts that no such stable identities exist, nor could such locations ensure the "perfect subject of oppositional history."[36] Because the subject is fragmented—never capable of occupying all subjugated positions or embodying any one such position—subjugation in itself cannot provide the "grounds for an ontology."[37] One may be subjugated, but by no means does this provide a guarantee that one will "see well."

Nevertheless, objectivity is not rendered meaningless according to Haraway. Instead, it comes to reside in communities constituted by "those ruled by partial sight and limited voice." These views interact and congeal to form a "collective subject position that promises a vision of the means of ongoing finite embodiment, of living within limits and contradictions."[38]

Constitutive criminology, especially the version articulated by Stuart Henry and Dragan Milovanovic, is also preoccupied with the problem of objectivity and viable ways of knowing. Henry and Milovanovic's *Constitutive Criminology* unfolds by comparing "modernist" and "postmodern" criminology. The latter can be broken down into its skeptical manifestations, such as a radical constructivist modality that restricts itself to deconstruction, and its affirmative versions, such as the constitutive criminology that Henry and Milovanovic develop. Despite this schism in postmodern thought, constitutive criminology is indebted to skeptical postmodernism, especially its tendency to focus on deconstruction. What it seeks to reject is the deconstructionist's tendency to slide into relativism. There is little reason to engage in *deconstruction* if this does not pave the way for *reconstruction*. The differences between these criminologies are revealed by examining how each conceptualizes the subject, social structure, law, crime, causality (of crime), and justice policy.[39]

Running parallel to the notion that society and science are always embedded in each other, discourse and practice are seen as interrelated for the constitutive criminologist. This is not to posit a unidirectional causal line that would link discourse to practice. It is not that discourse determines practice or vice versa, but that the two always coexist and remain mutually embedded. To encounter a discourse is to observe something that harbors practice; to observe practice is to encounter something shrouded in, and reproductive of, discourse.

Unlike the radical constructivist, the constitutive criminologist regards the interrelated nature of discourse and practice as a site for strategic intervention: if one is always found within the other, the creation of new discourses may engender new practices. Henry and Milovanovic thus argue for the importance of "replacement discourses," a notion that seems to place a premium on recognizing knowledge production as inseparable from political and ethical action.[40]

Henry and Milovanovic's analysis of "crime"—as a category that recurs across criminological thought—can be used to illustrate how "replacement discourse" operates in discursive practice. In most modernist criminology, crime tends to be understood as behavior that is caused by some underlying pathology and entails harm.[41] Much modern criminology asserts that crime can be explained by defects within the individual or the social, and that individuals, society, and/or the state are the entities most obviously victimized by crime.[42] This holds for modernists operating with a critical lens, who do not reject the idea that crime entails harm, but seek to expand the parameters of the victim category such that it includes "the underclass, women, racial and ethnic groups, sexual preference, employees, professionals, people with disabilities, people with AIDS, etc."[43] Despite these (and other) variations in the modernist camp, it is a view that ontologizes crime, effectively treating it as a signifier that embodies a real, objective activity.[44]

In skeptical postmodernism, crime and victimization do not exist outside of discourse. Crime, in other words, cannot be accorded ontological status; there is no direct line that allows us to construe specific practices as "criminal," as if some external reality "gave us" the category of crime. As shifts in power relations and discourse occur, the content that is poured into crime will change, intimating the latter's instability and arbitrary nature.[45]

The constitutive criminologist (and the modernist) is weary of this skeptical approach. If crime is only discursive, the category of victim loses any meaning and, moreover, allows those in power to deny that their actions can be assessed in terms of the harms that they may cause.[46] Consistent with constitutive criminology's promise of reconstruction, crime needs to be reconceptualized via this kind of skeptical deconstruction.

This ultimately leads Henry and Milovanovic to the notion that crime can be reformulated as "harm by reduction."[47] Whereas "crime" tends to subsume a fairly narrow range of behaviors that are codified in law, "harm by reduction" posits that the infliction of pain, suffering, and injury transpires within a broader context of power inequalities. Given this, crime is better understood as: "*some agency's energy to make a difference on others and it is the exclusion of those others who in the instant are rendered powerless to maintain or express their humanity. . . .* Crimes then are nothing less than moments in the expression of power, such that those who are subjected to them are denied their own contribution to the encounter and often to future encounters, are denied their worth, are simultaneously reduced and repressed in one or several ways."[48] Defined this way, "crime" or "harm by reduction" may exclude some practices that are currently criminalized but would criminalize various interactions that currently stand outside legal codes and are thus not proscribed.[49] Indeed, legal codifications of specific behaviors as criminal may in themselves amount to crime if such definitions exploit a power imbalance that represses. Moreover, as Henry and Milovanovic note, because "harm by reduction" presupposes power inequalities, it is necessary to alter the latter in order to address the former.[50] New practices would follow if this "replacement discourse" concerning crime were embraced. It would make little sense, for example, to restrict one's focus to retroactive forms of control that assume crime is constituted by isolated interactions in which a powerful subject reduces another to powerless object. While acknowledging the interactional component of harms, practice would focus on eradicating the power imbalances that are socially embedded—such as those grounded in race, class, sex/gender, sexuality, and so on—and exploited by some individuals, groups, and institutions in order to repress others.[51]

There are problems with successor sciences, situated knowledges, and constitutive criminology. While successor sciences and situated knowledges are distinct, both emphatically deny any possibility of occupying a location that allows a comprehensive view of the whole. Expressed in the terminology of situated knowledges, this is to perform a "god trick," to entertain the delusion that one has somehow secured access to a bird's-eye view of the world. Yet these positions, at least to some extent, claim this very space. How is it possible, one might ask, to know when a partial perspective, irrespective of whether it emerges from a subjugated location or a fragmented subject, amounts to objectivity from the margins or "seeing well"?

Perhaps more pressing, both positions may replicate one of the problems that C. Wright Mills identifies in his critique of "abstracted empiricism." As Mills points out, abstracted empiricism asserts that it can build a comprehensive picture of the objective totality by "stitching together" the particular studies and findings that its practitioners produce. However, the "who," "how," and "when" of combining studies seems to elude specification. That is to say, who recombines partial knowledges? What is the method for doing so? When is this reconstitutive work supposed to occur?[52]

What of constitutive criminology? In this affirmative postmodernism, its optimistic tendencies are perhaps best illustrated by the notion of "replacement discourse," which promises to provide a better framing of practice and thus a discursive strategy for reengineering practices that are seen as problematic. We use the example of "crime" and its reformulation as "harm by reduction" to illustrate Henry and Milovanovic's "replacement discourse." Here we see that the reality of "crime"—or harm—is not denied. Although Henry and Milovanovic state that skeptical postmodernism is correct in arguing that "crime is a socially constructed and discursively constituted category," they proceed to argue that it originates in defective social structures or power asymmetries.[53] It is hard to reconcile these two claims. As Carrier has pointed out, the second argument replicates the modernist criminology from which constitutive criminology wishes to escape or move beyond.[54] Of course, remaining trapped within modernist criminology is problematic, for it signifies that the constitutive approach does not offer a counter to many of the assumptions that underpin logical positivism. Instead of liberation from the tenets of positivism, it is representational epistemology, the quest to find some underlying sickness of which "crime" is a "symptom," a desire for intervention, and so on, that appears to be reproduced.[55]

Carrier's critique of "replacement discourse" (and, by extension, constitutive criminology) is compelling, but we are not utilizing it to surreptitiously assert the relative strengths of radical constructivism. The point we are working toward is quite different. It would seem that we are torn between epistemological positions that push the notion of discourse, and all the promise that it holds for critical analysis, to undesirable end points: if radical constructivism transforms the notion of discourse into a vehicle for venturing into relativism, successor sciences, situated knowledges, and constitutive criminology use it to make the return journey to "objectivity." These reworked conceptualizations of objectivity do break with logical positivism, but they often come dangerously close to remaining a satellite within its orbit. Are other destinations possible? Or, can the sense that one is obligated to travel be resisted?

Suspended Constructivism?

At the risk of offending the postmodern penchant for metaphors of movement, we wonder about the possibility of a "suspended constructivism." This would entail refusing the inclination to travel to either of the destinations promoted by radical constructivists or those attempting to reconceptualize objectivity. Perhaps it is best to remain in a state of tension between relativism and objectivism? This is not to suggest that one can unite these different positions via some kind of "synthesis." Instead, it is better understood as an attempt to work from their shared zones, or their points of overlap.

There are several ideas upon which radical constructivism, successor sciences, situated knowledges, and constitutive criminology converge. First, there is agreement that discourse and practice are always interrelated, or mutually constitutive. We use discourse and practice as broad signifiers or analogs for a range of co-constitutive polarities. In radical constructivism, the distinction between epistemology and ontology is collapsed; in successor sciences, it is science and society that are irrevocably intertwined, and the observer cannot extract their point of view from their social location. There is a "reality"—or something outside the subject—but it can never become "a-textual," escape language, or exist as an independent entity that allows itself to be apprehended in some pure, unmediated form.

Second, to posit that discourse and practice are interrelated provides a warrant for deconstruction, the critical interrogation of discursive regimes—especially those that purport to reflect reality without mediation. Deconstruction entails identifying and exposing the assumptions and contradictions by which texts are underpinned. Alongside this, it involves revealing the implications that specific discourses have for practice, the effects that the text produces.[56] By interrogating discourses in such a manner, the particular truths that they seek to establish as universal truth, and their corresponding practices, can be interrupted.

Third, radical constructivism and its critics appear to accept the proposition that drawing a distinction between discourse and power is untenable. It follows that interrogating any given discourse amounts to an interrogation of power relations. In this sense, constructivist approaches are politically engaged in some manner. To be sure, there are subtle variations on this point, which we have intimated in the preceding remarks. Radical constructivism may attempt to escape the political in its epistemology, but its politically engaged nature becomes difficult to deny the moment it offers substantive critique of texts. The refusal to ontologize crime, for example, is indebted to an antiauthoritarian stance that recognizes the impossibility of finding a legitimate basis for the exercise of state power. As Carrier puts it, what is the legal basis for drawing distinctions between legal and illegal acts?[57] Successor sciences, situated knowledges, and constitutive criminology strive for epistemological positions that are harmonious with the politically engaged knowledge that they produce or valorize. Despite the epistemological divides, all of these positions involve some recognition that critical practice or knowledge production inevitably partakes in political and ethical struggle, that something is at stake.[58]

But beyond this zone of convergences, additional theorizing leads us into the choppy waters of relativism or a reworked "objectivism." Marshaling the notion of discourse to argue in favor of a strong relativism undermines the very deconstructions that it enables. If all is relative, then any given critique cannot be accorded more weight than the object of critique. The return to objectivity disempowers by reasserting that some interpretations are "better" than others, that accounts can be judged for their capacity to reflect "reality" free of distortion (or with an appropriate type of distortion). Pushed to its extreme, objectivism returns us to a limited mode of debate in which interlocutors assume that they have access to "reality," one that is "beyond discourse" and thus unmediated. This runs the risk of reducing debate to a war of "facts," a struggle in which the disempowered will be severely disadvantaged. "Suspended constructivism" suggests that we operate from within the overlaps, that region of commonality in which discourse is recognized as a modality of power that necessitates critique. It is from this location that we interrogate crime science.

THE DISCOURSE OF CRIME SCIENCE

In the chapters that follow, we identify some of the fundamental branches that collectively constitute crime science, offering a critical reading of each. Before doing so, however, it might be useful to provide an overview of crime science as a discourse. We do this by exploring the work of Beccaria and Lombroso. Criminological textbooks often position them as founding figures of criminology and subsequently proceed to accentuate differences in their thought.[59] Of course, there is good reason to focus on points of disagreement: each offers a distinct way of conceptualizing "crime," the "causality" of crime, "effective punishment" strategies, and so on. As important as differences may be, the similarities in the thought of Beccaria and Lombroso are also worthy of exploration. For these similarities can be understood to constitute the broad parameters of crime science, and they reveal crime science to be a discursive arena that is quite capable of containing a variety of ideological positions. To put this otherwise, there may be variations, but they reside within the same discursive order. Our discussion of crime science as a discourse is organized around four themes. These include its ontology, epistemology, and fundamental categories; fetishization of technology; practical manifestations (especially in relation to law and criminal justice); and political allegiances.

Ontology, Epistemology, and Fundamental Categories

Crime science accepts that an order of reality exists independently of human meaning and practice. The objects with which it is concerned—human nature, crime, criminals, punishment—are assumed to belong to this external, independent reality. Such an account is, obviously enough, far removed from the constructivist positions outlined previously. Rather, crime science routinely accords its fundamental categories of thought ontological status.

Connected to this ontologizing of crime and other categories, one finds a particular epistemological stance. The detached observer can make sense of a reality that is assumed to be objectively given, carve it up into discrete variables, and lend it theoretical coherence by arranging variables into cause and effect relationships. By developing theoretical understandings that purportedly map this external reality, interventions or strategies for manipulating reality can be identified and implemented.

Although Beccaria and Lombroso do not utilize the same methods, it can be said that they are wedded to "representational epistemology."[60] Beccaria relies upon a philosophical mode of reasoning that deductively unfolds from the positing of "first principles." In Beccaria's case, it is the utilitarian maxim that pursuing the greatest good for the greatest number is morally desirable. This principle leads Beccaria to gauge the morality of legal, punitive practices that lend themselves to phenomenal apprehension and, insofar as they fall short of such a standard, suggest alternative strategies that would better embody the moral-practical implications of utilitarianism.

In Lombroso, the embrace of representational epistemology is much more obvious. Indeed, Lombroso has often been portrayed as responsible for transforming criminology into a "scientific" discipline.[61] Lombroso's belief that the world is distinct from its observation is readily discernible in his use of instruments that supposedly detect abnormalities in the human body. Such abnormalities, according to Lombroso, can be understood as symptomatic of underlying physical conditions that are conducive to crime. As is true of Beccaria, Lombroso is adamant that one can derive normative prescriptions—or what "should be done"—from what are assumed to be unmediated observations.

This ontological and epistemological domain is populated by the fundamental categories of crime science: human nature, crime, criminals, and punishment. According to Beccaria, human nature prescribes that individuals will be governed by self-interest and avarice and thus willing to sacrifice the collective good if it means securing private advantage. This does not necessarily mean that humans are irrational, but that their exercising of rational judgment will not go too far beyond an assessment of what is in their local self-interest. This cynical view of human nature is used to account for criminal acts, but also for those who err in the administration of criminal justice.[62]

Lombroso's construction of human nature is arguably much more convoluted than what is found in Beccaria. Lombroso maintains that human nature is defective insofar as it generates aggressive and violent behaviors among individuals. At some moments in human history, such aggressive tendencies were necessary for survival. But Lombroso holds that civil society no longer requires aggression and, as a result, demands its transcendence. The civilizing process, however, is uneven in its application; children are likely to follow their natural, depraved instincts toward aggression until properly socialized, and violent tendencies may persist through to adulthood within so-called civilized societies. This ultimately leads Lombroso to posit that some are capable of becoming "good," but others will always remain "evil."[63] Lombroso's "born criminal" embodies this latter notion, referring to individuals whose biological composition belongs to a past era of human history but resurfaces in the present by eluding the civilization process.[64]

Although both thinkers construe human nature as defective, they frame its undesirable outcomes in slightly different ways. In Beccaria, the troubling consequence that follows from the use of self-interested rationality is "crime," or the particular criminal act. For Lombroso, it is the "criminal," the individual with a propensity to commit criminal acts, that emerges as the core concern. Despite this difference, both views ontologize the categories with which they are preoccupied, effectively positing that "crime" (i.e., a particular act) and "criminals" (i.e., a criminally inclined person) can be taken for granted.

The idea that human nature harbors evil implies that distinctions between "bad" and "good," "criminal" and "noncriminal," are simple reflections of some supposedly natural, objective world. Rather than seeing discourse as the force that breaks the realm of human behavior into its legal and illegal manifestations, or the social body into its "criminal" and "noncriminal" factions, such distinctions are posited as readily discoverable within nature and thus self-evident. That criminality is regarded as a conceptual label that perfectly corresponds to behaviors found in nature is perhaps most evident in Lombroso's curious suggestion that animals display a variety of criminal behaviors, and that "ancient people were probably right to convict and punish animals that behaved dangerously or mutilated sacred objects."[65]

The distinction between crime and criminality is salient for unpacking how punishment figures in the discourse of crime science. Because crime or criminality is symptomatic of some defect in human nature (either the mind or body), it follows that punitive interventions intended to control crime should target the individual. For Beccaria, punishment is best utilized as a communication strategy that "educates" and thus deters people from criminal offending. Several conditions need to be met in order for punishment to function in such a manner: it must follow every crime, no matter how small; punishments must be calibrated according to offense severity and ensure that costs outweigh any possible benefits from crime; and punishment must never be based on emotional reactions, as this would lead to excessive penalties.

A system of punishment modeled on these principles would deter individuals from committing criminal acts, according to Beccaria. Those who have offended would learn that it is not in their self-interest to engage in crime again; punishments would deprive such individuals of the proceeds of crime and involve additional costs. In the case of those who are generally law-abiding, they would observe the operation of criminal justice and quickly recognize that punishment is guaranteed to follow crime.[66] While Beccaria's account posits that punishment is effective even if applied retrospectively—that is, as a response to crime—it is important to recognize that it also contains a preemptive logic. This is evident insofar as Beccaria holds that punitive practices have the capacity to communicate with the social body, informing it that violations of the law will be detected and censored.

Lombroso was much less concerned with the general deterrence that preoccupied Beccaria and thought that the function of punishment is to protect society from those with ingrained dispositions (evidenced by a variety of physical stigmata) that signal their high likelihood of engaging in criminalized behavior. Rehabilitating individual offenders and encouraging them to embrace rational decision-making—in short, striving to produce subjects that conform via moral training—are limited ideals, according to Lombroso. As such, punishment ought to serve the ends of prevention and incapacitation.[67] In this schema, it is the body that, in its very biology, is destined for a life of crime, thereby warranting its containment.

Crime Control as Technology Fetishism: "Operationalizing" Crime Science

Thus far we have seen nuanced conceptions of human nature, a tension between criminal acts and criminal dispositions, and different rationales of punishment. However, we have also seen how these variations stem from the same ontological and epistemic soil: human nature, crime, criminals, and punishment can all be understood as objects that reside within an external reality, one that is amenable to observation, thus "knowable" and open to intervention. A similar pattern can be observed when one considers how Beccaria and Lombroso seek to "operationalize" crime science; both figures posit that technological interventions are indispensable for the control of crime and/or criminality. Beccaria and Lombroso readily accept that a technical apparatus—incapable of thinking, acting on emotion, or pursuing self-interests—is the most viable way to identify and contain "human nature," thereby repressing its propensity for crime. This fetish for technology appears to follow, more or less automatically, from the postulate that human nature is defective: the machine or technical apparatus that does not function like a human cannot err like a human.

Although the details differ, the primary technological solutions advanced by Beccaria and Lombroso are identical in form: both are governed by a faith in syllogism. The syllogism that Beccaria posits effectively amounts to devising a formal, as opposed to substantive, system of justice. Following the view that no human can escape the pursuit of self-interest, Beccaria lays out a blueprint for determining punishments in a way that supposedly eliminates judicial discretion: "For every crime that comes before him, a judge is required to complete a perfect syllogism in which the major premise must be the general law; the minor [premise], the action that conforms or does not conform to the law; and the conclusion, acquittal or punishment."[68] In case there were any remaining doubt as to how circumscribed this exercise in logic ought to be, Beccaria hastens to add:

> If the judge were constrained, or if he desired to frame even a single additional syllogism, the door would be opened to uncertainty.
>
> Nothing can be more dangerous than the popular axiom that it is necessary to consult the spirit of the laws. It is a dam that has given way to a torrent of opinions.[69]

In light of this syllogism, it is not surprising to find Beccaria positing that a state's body of laws needs to be clear to everyone, encompass human behavior, specify which punishments follow from which crimes, and eliminate the need for interpretation.[70] One may sense that these kinds of demands are difficult to reconcile, if not contradictory; a body of laws that omits the need for any speculative interpretations and remains transparently clear to everybody seems a tall order. Nevertheless, for Beccaria the utility of the syllogism resides in the fact that it eviscerates the possibility that subjective assessments by judges will interfere with the task of meting out state-sanctioned punishments. In the scenario proposed, all the judge needs to do is ascertain the "facts" of a case (as if this were straightforward) and subsequently classify behavior according to a written body of law. If an act constitutes an instance of what is forbidden by general law, a prescribed penalty is imposed. This mechanistic view is not some quaint artifact of the past. It is a view with contemporary relevance, still held by some leading jurists. Indeed, in the opening remarks of his 2005 confirmation hearings, John Roberts, chief justice of the United States Supreme Court, explained: "Judges and justices are servants of the law, not the other way around. Judges are like umpires. Umpires don't make the rules; they apply them. The role of an umpire and a judge is critical. They make sure everybody plays by the rules. But it is a limited role. Nobody ever went to a ball game to see the umpire."[71]

In Lombroso, Beccaria's "major premise" of a general body of laws is reconfigured as a typology of general criminal types; instead of an isolated act, the "minor premise" is an individual offender. The prescribed mode of punishment depends on how an individual is classified. Rather than a judge executing the program, the criminal anthropologist—assumed to be an objective and impartial scientist—ensures that the classification process is logically coherent.[72] As in Beccaria, this syllogism is fetishized for its promise to remove human subjectivity and fallibility from judgment.

Unpacking Lombroso's general law, that is to say his typology of criminals, is complicated by the fact that *Criminal Man* went through five editions between 1876 and 1897, with new editions marking shifts in Lombroso's thought. By the final edition of *Criminal Man*, Lombroso seems to suggest that four broad criminal types exist—the born criminal, the insane criminal, the morally insane criminal, and the criminal epileptic—some of which contain subtypes. Lombroso is adamant that criminal types can be identified by physical and psychological traits, although these will be more or less discernible depending on the extent of one's supposedly innate criminality.

The born criminal is, without doubt, Lombroso's most well-known category. In defining born criminals, there is a repetitive emphasis on physical and mental traits. As Lombroso sums up the category:

> [C]riminals resemble savages and the colored races. These three groups have many characteristics in common, including thinness of body hair . . . small cranial capacities, sloping foreheads, and swollen sinuses. Members of both groups frequently have sutures of the central brow ridge, precocious synostes or disarticulation of the frontal bones, upwardly arching temporal bones, sutural simplicity, thick skulls, overdeveloped jaws and cheekbones, oblique eyes, dark skin, thick and curly hair, and jug ears. . . . In addition, in both we find insensitivity to pain, lack of moral sense, revulsion for work, absence of remorse, lack of foresight (although this can at times appear to be courage), vanity, superstitiousness, self-importance, and, finally, an underdeveloped concept of divinity and morality.[73]

As this passage intimates, Lombroso's description of the born criminal may serve equally well as a succinct summary of the most well-worn racist stereotypes within Anglo-Saxon and white-European societies.[74] In any case, Lombroso accounts for born criminals by suggesting that they represent atavistic throwbacks to earlier moments of human evolution. He also suggests that born criminals are a product of arrested development or disease, including particular forms of epilepsy.[75] Given these biological origins, it follows that born criminals will inevitably engage in criminalized behavior and that they are beyond reform. The most suitable punishment is thus permanent incarceration, and the death penalty for repeat violent offenders.[76]

At the other end of the spectrum, Lombroso identifies criminals of passion as those who violate the law due to strong but anomalous

FIGURE 2. Lombroso's gallery of delinquents. Several features—clear partitions between subjects, the presumption that criminality can be seen or visually detected—resurface in brain scan imagery (cf. figure 4). *Source: L'uomo delinquente in rapporto all'antropologia, alla giurisprudenza ed alla psichiatria.* Atlante/Cesare Lombroso, 5th ed. (Torino: Fratelli Bocca, 1897).

outbursts of emotion. These criminals display few physical or psychological abnormalities, and it is unlikely that they will re-offend. As such, Lombroso suggests that they are not in need of severe punishment, and sanctions such as fines and "judicial admonition" may be appropriate.[77] Like other categories proposed by Lombroso, the passionate criminal is contemporaneously recognized in various respects. The label refers to those who commit a serious crime but can successfully frame their behavior as "out of character."[78]

The technical solutions to "crime" and "punishment" that Beccaria and Lombroso propose are fetishistic; they involve promising that "scientific-technical" interventions will magically fix the broader problems that otherwise populate the discourse of crime science, if not bring to fruition some image of the perfect society. This technological fetish often leads to irresolvable contradictions.

The proposal to turn the punishment process into a formal system, one that is insulated from subjective desires and interests, is construed by Beccaria as central to producing a better society. A clearly codified body of laws that are impartially applied will prevent crime, and crime prevention is intimately tied to producing the greatest possible happiness for the greatest number of people.[79] Transparent and systematic legislation will create common understandings of appropriate behavior and impress upon a society's members that violating formal rules will have adverse consequences for one's personal happiness. In short, Beccaria sees formal justice as fulfilling the utilitarian maxim that society ought to pursue the greatest good for the greatest number.[80]

Beccaria's promise of happiness via formal justice, however, obviously hinges on a massive criminal justice apparatus capable of identifying and processing every infraction of the law. This, in turn, entails that formal law will draw a circle around human behavior and subject everyday life to perennial surveillance. The demand on the superego to internalize this proliferation of law and regulate behavior accordingly sounds taxing, to say the least. Insofar as law hinges on the use of force, the promise of happiness would appear to be betrayed: to embed contentment so thoroughly within the force of law is to produce an image of dystopia.

In acknowledging that the eradication of crime is unlikely, but that there may be some possibility of preventing and "curing" criminality depending on what type of offender the justice system is encountering, Lombroso seems a little more circumspect in outlining the benefits of embracing his typology.[81] His account entices the reader through a promise of security. Against this, however, he also claims that the practical implementation of his suggested policies, especially those punishments intended to address born criminals, may facilitate human evolution: "To claim that the death sentence contradicts the laws of nature is to feign ignorance of the fact that progress in the animal world, and therefore the human world, is based on a struggle for existence that involves hideous massacres. Born criminals, programmed to do harm, are atavistic reproductions of not only savage men but also the most ferocious carnivores and rodents. This discovery should . . . shield us from pity, for these beings are members of not our species but the species of bloodthirsty beasts."[82] In construing the death penalty as a method to accelerate the pace of human evolution, Lombroso would appear to give new meaning to the notion of hubris. His more serious promise, however, is greater security through a mix of prevention, cure, and incapacitation. The emphasis on security leads Lombroso to make numerous suggestions that, on the surface, sound quite critical. There is, for example, the acknowledgment that prisons are a breeding ground for further criminal activity and therefore their use ought to be minimized.[83] There is also the suggestion that governments have an obligation to address economic inequalities and ensure that work is available so that people can meet their everyday needs. In some sense, Lombroso even hints at the possibility that crime is a constructed category, suggesting that a wise nation would consider channeling the energy that typically goes into criminalized activities into more socially desirable pursuits, such as sports or circus performances.[84]

However, racist ideologies are inseparable from Lombroso's typology of criminals, making it hard to accept that he is promoting socialist policies, rehabilitative forms of punishment, and universal security. We would say that the criminal types demarcated by Lombroso amount to a barely veiled repository for racist sentiment, evidenced by the fact that criminality magically seems to dissipate for offenders with weaker concentrations of melanin. At best, this is "socialism" and progressive punishment for those effectively perceived as racially superior. Those not displaying enough "European features" cannot be helped, and their lot in life is perennial detention or death. This amounts to an incredibly insecure world for those who possess certain physiological features.[85] For those who possess only some of the physical traits that Lombroso considers signs of criminality, the risk of being subject to surveillance and behavior modification programs means that things are only moderately better.[86] The few people remotely secure in this world would appear to be those with "anticriminal faces with broad foreheads, rich beards, and gentle, serene countenances."[87]

Instantiated Practices of Crime Science: Law and Criminal Justice

Many features of criminal justice (in the Anglophone world) can be read as embodying the crime science espoused by Beccaria and Lombroso. Consistent with the "first premise" of Beccaria's syllogism, for example, modern criminal justice is premised upon a proliferation of legislation that seeks to delineate, in ever-greater detail, the borders between illegal and legal behavior. One can also discern, though arguably to a lesser extent, a growing body of written laws that specify formal procedures governing the operation of criminal justice. This is manifest in legislative acts that regulate police powers and the relatively recent emergence of sentencing guidelines. These latter aspects of legislation would appear consistent with Beccaria's warning that if too great a degree of discretionary power is extended to agents of criminal justice systems, it is likely to be misused and thereby generate injustices.[88]

Of course, by no means does this expansion of written laws that seek to construct a formal system of justice guarantee that society's members will be clear about the rules that effectively govern their lives. If anything, the opposite seems to be true: it is incredibly difficult to follow rapid developments in legislation. In fact, it has been noted that in the United States there are so many federal laws, scattered across different legislative codes, that "no one actually knows how many there are."[89] Moreover, written laws rarely make for clear, intelligible reading, as Beccaria had hoped. Likewise, efforts to minimize the discretionary exercise of power through legislation do not seem capable of resolving the kinds of problems that Beccaria identified, such as inconsistent application of penalties.[90]

At a deeper level, the collective investment that lurks behind an ever-expanding body of legislation mirrors Beccaria's faith in the power of formal laws to produce a better society. It would seem that many readily accept the proposition that criminalization, innovative penalties, and new arms of criminal justice ought to enclose the social. This is perhaps most evident in the rise of specialized courts, such as those intended to address drug and alcohol abuse, youth offending, homelessness, and family violence.[91] Such courts obviously have some connection to behaviors that have been criminalized, but they also indicate the way in which criminal justice is increasingly tasked with intervening in what are often held to be social problems.

The "minor premise" of Beccaria's syllogism—does the particular act conform to or violate law—presupposes that the reality of crime can be established. Thus the quest to discover techniques and strategies to prove that a crime has occurred and can be attributed to an individual. In his analysis of how the body has been compelled to produce truths about crime and criminality, David Horn draws a distinction between "signaletics" and the "semiological."[92] Signaletics is governed by a desire to prove who committed a particular act of crime; the semiological strives to identify the "essence" of individuals, to discover who is biologically predisposed to engage in crime. The most well-known manifestations of signaletics are, arguably, the fingerprint, DNA profiles, and the "lie detector." We are not suggesting that Beccaria is the "founding figure" of signaletics. However, we do suggest that a logical consequence of Beccaria's emphasis on locating acts within a general body of law provides a warrant for seeking out technologies that purportedly connect the individual to the criminal act.

Concerning the legacy of Lombroso, it is not much of a stretch to say that the most pernicious elements of his thought are encoded within various criminal justice systems, and that this is evident in their practices and outcomes.[93] "In every country, crime and incarceration rates for members of some minority groups greatly exceed those for the majority population."[94] This is especially true of countries that are hierarchically organized along racialized lines and those with colonial pasts and presents. Incarceration rates in jurisdictions such as the United States, Australia, and New Zealand demonstrate that racially marginalized individuals are overrepresented in prison populations. Incarceration rates are also very high in such places, giving rise to the notion of "mass incarceration."[95]

There are various drivers of racialized incarceration, but the phenomenon is undeniably accompanied by the logic of risk assessment entering the sentencing process; commonly held views that rehabilitative efforts are useless, and thus harsher punishments are necessary; and, among other things, acceptance of the idea that the primary function of prisons is to incapacitate offenders.[96] These ideas all echo positions advocated by Lombroso and, when they play out in contemporary contexts, subtly absorb the racialized dimensions evident within his construction of criminal types.

Lombroso's efforts to locate criminality in the human body via specific measurement techniques or devices are often discredited and ridiculed, but like ghosts that do not want to be exorcised, they also continue to haunt criminal justice.[97] To return to Horn's work, he shows that criminal justice and criminology remain invested in "scientific instruments" that are thought capable of revealing the "essential nature" of individuals. In Horn's terminology, this is the "semiological" side of compelling the body to act as a site of truth about "criminality." Examples of such technologies include the penile plethysmograph, which measures "the flow of blood to the penis" and is used to identify sex offenders, but the tendency is also evident in the study of brain scans of psychopaths (and others) and in efforts to "'decode' and 'read' the human genome."[98] The latter examples are consistent with Lombroso's view that bodily properties can be measured to reveal tendencies toward criminality: the brain scan is held to reveal "brain activity patterns" that correlate with violent crimes of passion, and the genome is held to reveal genetic markers that correlate with violent inclinations and behaviors.[99] Interest in the "psychology" of criminals may also be taken as evidence of the ongoing salience of Lombroso's view that criminality inheres in the body. The contemporary view that criminalized behavior stems from "cognitive distortions," for example, essentially posits that criminals somehow "think differently" than the rest of us, thereby generating crime.[100] The weight that Lombroso gave to "moral sensitivity" and the emotions is an obvious precursor to this line of thought.

Other elements of contemporary justice appear to be a product of the tensions—arguably irreconcilable—between Beccaria and Lombroso. Following Beccaria, few criminal justice systems, if any, have rejected the view that individuals can be held accountable for their actions. Consistent with Lombroso's questioning of responsibility, however, there is a tacit recognition of situational forces that constrain choices, thereby mitigating criminal culpability. For example, courts hear defenses based on duress, self-defense, and necessity; sentencing judges consider factors such as youth, illness, and prior abuse. The tension between causal (positivist) factors and agency (free will) is rarely disentangled in criminal justice settings. Herbert Packer explains, "Very simply, the law treats man's conduct as autonomous and willed, not because it is, but because it is desirable to proceed as if it were."[101]

In terms of punishment, many criminal justice systems simultaneously embrace an array of punitive techniques, even though these may be underpinned by disparate rationales. Imprisonment and penalties that exceed this, such as preventive detention and the death penalty, are broadly consistent with Lombroso's argument for sentences that incapacitate. But sentencing hierarchies also make use of community detention, community work, supervision/probation combined with rehabilitation programs, fines, and good behavior bonds.[102] This would appear to follow from Beccaria's view that punishments need to be calibrated to different crime types and focus on (re)engineering how subjects calculate possible courses of action.

Political Allegiances: State Authority and Power Asymmetries

One may discover critiques of particular criminal justice policies and practices within Beccaria and Lombroso, but the fundamental interests of the state are not challenged by their work. If anything, Beccaria and Lombroso are united by their tendency to construe the state and the power arrangements within which it operates as legitimate.

Beccaria seeks to develop punishment logics that ensure tighter regulation of the population. Moreover, the population control desired by Beccaria would function by ensuring that subjects self-monitor their behavior, thereby minimizing the need for an external source of authority that is overtly dependent upon the use of force. This is evident in Beccaria's call for a body of laws that exhaustively documents proscribed behaviors and that should become general, common knowledge. As Foucault put it, along with other reformist discourses of the period, Beccaria is not necessarily interested in creating humane punishment, but in "punishing better." The goal is to create a legal apparatus that pushes regulation deeper and deeper into the social body.[103]

It is noteworthy that Beccaria negates the possibility that judges can be entrusted to exercise discretion without introducing unpredictability in sentencing. This is often read as evidence of Beccaria's concern to rationalize criminal justice, such that law's consistency is assured. But in light of Foucault's reading, the seeming desire to reduce judges to mechanical applicators of law takes on new meaning: it is to compel the imposition of law as it is written. In this, Beccaria appears to assume that written law "falls from the sky," that it can somehow escape mediation by political interests.

This is odd in light of the claim that human nature is defective, driving individuals to pursue private interests; while judges cannot resist this defect, legislators are somehow immune to it. Written law may facilitate consistency in its application, but this does not mean it is free of political biases. It might just mean that law is consistently skewed in a particular manner. The political import of Beccaria's presumption of an "immaculate" body of law, one that is mechanically applied, becomes clear when one notes that it has obvious parallels with conservative demands for "truth in sentencing," "law and order," and legislative codes that are "tough on crime"—all of which erroneously assume that law can be produced in a manner that does not encode political positions concerning crime and punishment and thus can be mechanically applied in politically neutral ways.

Lombroso also sought to push the state's punishment apparatus in new directions but did not appear to have much interest in questioning the state's legitimacy or its power to regulate populations. In the first edition of *Criminal Man*, Lombroso went so far as to say that theory should not proceed beyond a point at which it becomes disruptive: "Nothing is less prudent than the attempt to carry theories to conclusions that could disrupt society... Fortunately, my scientific findings, far from making war on social order, reinforce it."[104] This remark was a response to criticisms of Lombroso's view that many forms of crime are biologically determined. Critics reasoned that if crime was beyond the control of an individual, then it was unjust for the state to punish.[105] Lombroso would sooner plug the excessive meanings within his theories on criminalized behavior, truncating any practical implications that are perceived as undesirable, before allowing them to interfere with state punishment. If anything, Lombroso was invested in intensifying the state's power to punish: "While we may deny individual responsibility, for it we substitute social responsibility, which is much more exacting and severe. While we may reject the legal responsibility of criminals, we do so not to reduce their punishment, but rather to increase the length of their detention."[106]

The individual is not responsible for their actions, then, but the social is responsible for guarding against those deemed biologically defective. In Lombroso, protection from bodies that harbor danger rationalizes preemptive strategies, severe punishments, and sustained periods of confinement.[107] To refuse this task becomes immoral, socially irresponsible. The state is entrusted with this power to confine, and because it is assumed that some greater social good is secured by its exercise, such power is rendered legitimate.

While Lombroso recognizes that social relations are plagued by a conflict, this does not revolve around power asymmetries grounded by "race," class, gender, sexuality, and so on. Instead, the fundamental conflict involves an antagonism between those who are "law-abiding" and those whose biological make-up assures their "criminality." But, given that Lombroso's typology of criminals—from the "born criminal" to the "criminal of passion"—is suffused by racist ideology, positing the fundamental social conflict as a struggle between the "law-abiding" and "criminal" is disingenuous, a barely veiled strategy to recode and thus normalize racial hierarchies. In sanctioning the constraint of those whose biology destines them for a life of crime, it is a racialized body that will ultimately find itself quarantined. Of course, "race" is only one axis along which contemporary power asymmetries are organized. Nevertheless, Lombroso's uncritical acceptance of state power, combined with a theory of criminality suffused by racial ideology, is indicative of crime science's political allegiances. Crime science legitimizes state power and its utilization such that social power asymmetries are preserved. Lombroso attempts to portray state power as the guarantor of *general* social security. Beneath such rhetoric, however, it is the security of *particular*, privileged groups that is pursued by forcing "others" to sacrifice it.

Harding has argued that particular political agendas pass straight "through the scientific process to emerge intact in the results of research as implicit and explicit policy recommendations."[108] And, in a slightly more provocative formulation, "science is no different from the proverbial description of computers: 'junk in; junk out.'"[109] This occurs because social relations and science are so intertwined. The point neatly encapsulates crime science.

According crime an ontological status (treating it as a category that is beyond question), constructing it as a symptom of some defect in human nature, and regarding it as a "technological" problem is the "junk" that goes into crime science. The "junk" that comes out is a fetish for detection of crime or criminals, mechanical justice, containment or repression of (specific) bodies, and moral conditioning of the population. But this is to construct a problem in a very specific way. One could quite easily put a different kind of "junk" into crime science, thereby ensuring that something else emerges on the other side. If, for example, "crime" is understood as symptomatic of a defect in economic relations, then socioeconomic solutions are required. Or, adopting a radical constructivist stance, if crime is understood to consist in "the effective use of the difference criminal/non-criminal," then the output is a renewed logic for the "critique of dominating and authoritarian practices."[110]

Despite pretensions to the contrary, then, crime science does not amount to a "value-free" or "objective" scientific account. As the first section of this chapter suggests, the very notion that an "objective account" is possible presupposes a disembodied standpoint that does not exist. Crime science presents itself as disinterested, an unmediated mapping of reality, but is better understood as a discourse that cannot be dissociated from state authority and a range of social power asymmetries. This is evident in the categories that it regards as self-evident, its arrangement of those categories into cause-effect relationships, its fetish for punitive interventions and social control, and so on.

Faith in the discourse of crime science is still with us today. If anything, it operates as a dominant discursive regime. Its central categories—defective individuals, crime, criminality, punitive control—spawn a seemingly endless range of scientific-technical interventions designed to suppress behaviors constructed as criminal. It is these discursive and technological interventions, often embraced by the state, law, and criminal justice, that the following work critically interrogates.

CHAPTER 2

Biological Crime Science

Identification and Biosocial Criminology

At first glance, biological crime science appears to consist of at least two branches: technoscience and academic discourse. Technoscience concerns those moments in which, as Andrea Quinlan suggests, science and technology are collapsed.[1] In relation to biological crime science, this would roughly translate into various forms of forensic evidence, such as fingerprinting and DNA typing, which rely on the body's properties to help to determine whether an individual is guilty (or not guilty) of a criminal act. Academic discourses in this domain correspond to biological criminology and its revival in what has come to be known as *biosocial criminology*. Such discourses are generally on a quest to develop causal theories of criminality, in which forces that supposedly inhere within the individual are accorded a central place.

This neat division, however, is somewhat illusory because bio-forensic technologies and biological criminologies have much in common and often flow into one another. In terms of common ground, both strands accord crime an ontological status and see it as something in need of resolution. Moreover, and insofar as they are understood as technologies of identification, fingerprinting and DNA typing facilitate the development of criminal records, which catalog the offending behaviors of an individual. Criminal records can be, and often are, subsequently taken as evidence of "criminality." The more extensive one's criminal record, the more readily they can be referred to as a "habitual offender," "career criminal," "life-course persistent" offender, or, perhaps more commonly, "recidivist." As Simon Cole and Michael Lynch and colleagues point out, the development of criminal records—a result of biometrics that identify individuals—has spawned innovations in criminal justice practice that target "criminality." Many states now punish for recidivism in general and intensify penalties for repeat violations of particular penal codes.[2] Punishing recidivism transpires on the assumption that criminal justice systems can, and should, cleanse one of their "criminality."

It is not necessarily the case that bio-forensics emerged, then subsequently inspired biological criminologies. The two branches have long been self-reinforcing; the idea of innate criminality is better understood as coterminous with the use of biometrics in legal contexts, and the belief that biological markers hold predictive value has often run parallel to their incorporation into criminal justice settings. As Cole notes, Frederick Brayley thought that fingerprints were "part of the plan of the Creator" to eliminate crime because it was thought they could be used to detect criminality in individuals. Likewise, Francis Galton saw fingerprints as an "identifier and a hereditary marker."[3] In the present moment, very few would take seriously the suggestion that fingerprints can identify those with "innate" tendencies toward "criminality," thereby separating them from those who harbor no such inclinations. However, the suggestion that DNA or genetic material, brain patterns, psychological testing, and so on can locate criminality somewhere within the body would not seem a far-fetched notion to many people.[4]

This chapter concentrates on three themes that circulate in the area of biological crime science. First, it considers the tendency to construct bio-forensic technologies as capable of resolving crime. We suggest such technologies are better understood as dependent upon crime. At their center one finds an irresolvable contradiction: a promise to solve the problems that sustain their very existence. Moreover, bio-forensics actively produce crime, conjuring it into being in various senses. As Corinna Kruse notes, within formal legal settings bio-forensic evidence plays a pivotal role in categorizing events as criminal.[5] However, bio-forensic technology also works to reframe everyday behaviors—especially those that are legally protected—as deviant or as evidence of lurking criminality.

Second, the chapter explores the notion of bio-evidence as an infallible, unassailable discoverer of truth. Infallibility is often attributed to DNA profiling, but a similar sentiment has engulfed other types of bio-evidence, such as fingerprint examination. Scholars operating within science and technology studies paradigms have heavily critiqued claims of infallibility, revealing how bio-forensics is open to error. As such, there is reason to be skeptical of bio-evidence and its purported ability to ensure certainty in legal judgments, thereby guarding against false convictions.

We agree with this skepticism but explore a slightly different problem in terms of infallibility. In our view, one of the main effects of bio-forensics entering criminal justice has been to promote differential standards of legal truth and calibrating these to a hierarchical ordering of crime types that bio-forensics takes for granted. The relatively few crime types in which DNA evidence is usually available—murder, some forms of sexual assault, and some property crimes—come to be seen as leaving little room for error. However, in those areas where DNA evidence has little bearing, which is the case for most forms of offending, the threshold for certainty will be much lower. In other words, rather than turn criminal justice into a scientific enterprise, DNA evidence (and by extension, other bio-evidence) skews our perceptions of when relative certainty is required for establishing guilt.

Third, we switch gears and examine biosocial criminology, arguing that the problem of equivocation engulfs many of its fundamental categories. We begin by analyzing how "crime" and "victimization" are defined in incompatible ways, and how different definitions are used depending on who is being spoken about. Those from racially marginalized social groups are routinely rendered criminal, whereas dominant groups may engage in behavior that is "horrific"—such as the orchestration of slavery or genocide—but somehow elude being cast as criminal. Similarly, the "victimization" experienced by marginalized groups is portrayed as a psychological deficit, if not a delusion, but victimization is presumed to be "real" or "objective" for others. Following the analysis of "crime" and "victimization," we look at three other constructs that are treated equivocally—the science-society relationship, the meaning of numbers, and "adaptation"—and show how the political standpoint of biosocial criminology accounts for these equivocations.

Throughout the chapter's exploration of the contradictions that haunt bio-forensics and biosocial criminology, attention is drawn to the politics by which biological crime science is animated. In this respect, we focus on how biological science inscribes criminality on those who are compelled to live life at the margins of society and how this exonerates contemporary power relations.

PROMISES OF CRIME RESOLUTION, OR "HOCUS POCUS" BULLETS

Technologies that emerge at the intersection of the body and science are often construed as "magic bullets" that will resolve the "crime problem." The most obvious promise of biological technologies stems from the belief that things like fingerprinting and DNA typing can discover the authors of criminal acts.[6] The body—its oily residues, fluids, and matter—is thought to leave a double of itself at a crime scene. It is this body double, the traces left behind, that can be detected and used to append the criminal act to the individual. Once the criminal act is connected to its individual author, the various branches of criminal justice can be set in motion to ensure that an offender is punished. The imposition of punishment is thought to reduce crime.

A second type of promise—one in which bio-evidence is read for its potential to deter—is also made. Here, a general awareness of the body's doubling capacity and, more specifically, the view that the body cannot but give itself away, will ultimately resolve crime. In such a view, it is assumed that one cannot contain their body and its properties: a fingerprint will be left somewhere; a hair will fall out in the course of committing a crime; sweat, saliva, or blood will be deposited. These samples will be searched against fingerprint and DNA databases, which are already massive and continue to grow. Repeatedly searching crime scene samples against expanding databases means that one may escape detection for some time, but eventually bio-forensic technologies will identify and ensnare the individual offender. The belief that the body cannot be outrun figures as the ideal embodiment of deterrent logic: there is no escape from the machinery of bio-evidence; you will be caught. In this promise to end crime via its deterrent capacity, bio-forensics figures as an omnipotent god, capable of seeing all and never failing to arrive at a moment in which error-free judgment will be passed.[7]

Like most "magic bullets," however, these promissory notes turn out to be illusory. In saying this, we do not mean to imply that biological forms of evidence are deficient because they cannot actually resolve "real" crime problems. In other words, we reject the view that crime can be accorded an ontological status, then be treated as a problem that is amenable to some kind of technical solution. Bio-forensics, however, certainly construes crime as an objectively given reality, then positions itself as instrumental to crime's resolution. In making such assumptions, some of the logical contradictions inherent to bio-evidence become discernible.

Several critiques hinting at the self-defeating nature of bio-evidence have been made. In studying forensic work in Sweden (tool mark analysis, fingerprint comparisons, DNA typing), Kruse suggests that such techniques do not simply map the reality of crime, but work to successfully construct acts as criminal and individuals as guilty of criminal acts.[8] This argument follows from analyzing how things like fingerprints and DNA samples must make their way through complicated legal systems consisting of multiple agencies, each with its own objectives and "epistemic cultures."[9]

Any given DNA sample, for instance, is sourced at a crime scene by a police investigator. From here, it is passed to a laboratory for analysis and, possibly, turned into a sample for which there is a match in a database. The sample then travels to a prosecutor, who does an interpretive reading and takes both into a courtroom. But the interpretive reading may well encounter an alternative narration provided by a defense lawyer. At the end of this chain, the DNA sample and its supposed double in a database will be accepted as part of either a narrative that establishes the guilt of the accused (prosecutor's version of events) or a counternarrative in which it fails to establish guilt (defense's version of events).

From this, Kruse draws an important conclusion: DNA and other forensic evidence never speaks for itself; it is never reducible to a simple signal that is transmitted to us from the "real" world. Rather, it is embedded within competing narratives that strive to establish guilt or innocence. As Kruse states: "The verdict the court makes on a case is an interpretation of the available evidence. This also means that the court is less of a fact finder and more of a fact *maker*. . . . [R]eality is the outcome of settling controversies, not the cause of the settlement. . . . Defendants are not convicted because they are offenders, but they become offenders through being convicted. The court does not discover the truth; rather, it makes the truth."[10] The forensic evidence that is supposed to eradicate crime does the opposite. For it is this very evidence—along with other legal processes—that is central to establishing crime. This argument cannot be dismissed as semantic; it is not about describing an otherwise stable reality with a peculiar linguistic repertoire. Instead, the bio-forensic machinery is dialectically related to crime, the latter being central and indispensable to the former. Crime is the very thing spun from bio-forensics.

Focusing on the use of DNA technologies in the Canadian legal context (primarily databanks and matching techniques), Neil Gerlach also posits that they cannot be understood as a simple effect of the desire to dissolve crime problems. In his view, DNA technology is better understood as a "postpanoptic" risk management strategy.[11] Rather than eradicate crime, such technologies identify and sort populations into "high" and "low" risk groups, then subject these groups to different levels of surveillance and social control. As Gerlach writes: "DNA banking serves as a method for coding an individual as part of either a network of inclusion or a network of exclusion. It is one means among several for classifying people according to the level of risk they pose to circuits of civility. It serves as a gate-keeping mechanism to separate the responsible from the risky, the seduced from the suppressed."[12]

The imperative to surveil and separate is "postpanoptic" because it is not about correcting individuals in the sense suggested by, for example, Foucault.[13] DNA technologies operate minus any ambition to move people from high-risk status to low-risk status; they simply quarantine (temporarily for the most part) those deemed high risk, thereby leaving intact the underlying cartography of risk that they enact and reflect. For Gerlach, risk management comes at a cost insofar as it sacrifices individual rights and protections from undue state interference for promises of greater public safety and enhanced crime control.[14]

We do not disagree with these theorizations. Rather, we wish to add some further lines of critique that follow from the interfacing of bio-forensics and criminal justice and the associated promise to resolve crime. Three other critiques are worth making in our view. First, bio-forensics harbors two logical problems or contradictions. Second, it rests upon a fallacy that we refer to as the "immaculate database." Third, not only does bio-forensics squash individual rights, it turns any exercise of such rights into social deviance and thus grounds for suspicion. To treat the promises of democracy and its foundational documents with reverence is pathologized within what Gerlach refers to as the "genetic imaginary."

Two Logical Contradictions

In terms of logical problems, the first stems from the dependency of bio-forensics on what it construes as crime problems that ought to be eradicated; it is as if bio-forensic technologies must lie in wait for "crime," like a hunter that needs the woods to be populated with deer, not emptied of them.

Quinlan opens her article "Technoscience and Affected Bodies" with a quote from a narrative by a survivor/victim of rape. It is worth reproducing at length because—in addition to demonstrating (as it does for Quinlan) the violence of rape kits and their capacity to revictimize individuals—it also encapsulates the problem we have suggested:

> My number was called and I'm wheeled into a medium-sized room where three women in white lab coats are waiting. They speak softly, tell me their nurse and doctor names, [and] what they will do. . . . Then they ask me to tell my rape. *They write everything down, record data on forms with numbers and codes that have been waiting for me to be raped*. . . . They are efficient and distant as they spread a circular plastic sheet on the floor and ask me to stand on it and remove my robe. They brush the hair on my head and between my legs pluck[ing] fifteen pubic hairs by the root . . . stirrups, gloves, stainless steel inside me, entering, expanding. . . . Everything is collected in vials and plastic or under glass, labeled with my name. All these pieces of me are placed in a kit to be touched and examined, probed and considered some more, somewhere, by someone, for something.[15]

In this passage, the author makes direct reference to the strangeness of a technoscience that has been developed and organized in anticipation of victimization, yet is incapable of actually preventing that victimization—the forms, numbers, and codes *await* rape victims, assuming that they will be making their way into the laboratory. In addition to the references that are more or less direct, the passage alludes to the logical contradiction of bio-evidence in other ways. Hairs are brushed onto the plastic sheet for the sake of preservation and analysis, and a range of vials and containers will also house various materials collected from the body. These constitute the "rape kit," which will be sent off somewhere for further analysis.

The "rape kit," with all its various parts, is manufactured by companies, and there are laboratories that are built and arranged such that kits can then be analyzed, matches searched for, and so on. Not only, then, is there a "rape kit" that sits on a shelf waiting for victims, but there is also a vast infrastructure that is necessary for it to "make sense," for its legal significance to be drawn out. This infrastructure is a complex assemblage of spaces, people, technologies, expertise, and social institutions.

Of course, this victim's account traces an individual experience, but it has obvious bearing on the limits of bio-forensic technology broadly conceived. We argue that most, if not all, forms of bio-evidence utilized in contemporary criminal justice systems assume that criminalized behavior can be expected but not prevented. And yet it is simultaneously promised that bio-forensics—especially if we make greater investments in it, build bigger databanks, expand its operation into more and more areas of criminal justice—will solve crime problems. The problem, however, is that expanding databases and the use of bio-forensics does not make sense if one is assuming the eradication of crime as the ultimate telos toward which the machinery is directed.

The second logical problem is buried within the promissory notes of bio-forensic technologies, which assume two possible types of criminal actor. Neither of these actors is understood as a product of social contexts; bio-forensics does not proffer "structural solutions" or call the category of crime into question as constructivists might. Rather, both actors are presumed to be governed by forces operating within the individual. The promise of deterrence assumes a rational actor, one that will see in bio-forensics the certainty of punishment and thus desist. The promise to identify and punish criminal acts retrospectively assumes an irrational or miscalculating actor; the technology exists, but it somehow does not deter.

But presupposing a criminal actor that is either rational or irrational cannot be easily squared with the promise to eradicate crime. If a rational actor is posited, it is just as likely that steps will be taken to evade detection; if an irrational actor is assumed, then crime will occur anyway, given that it is driven by forces that transcend legal codes and the state's capacity to punish. On this point, it might be worth recalling Jack Katz's notion of "seductions of crime," in which many crimes are understood as projects to restore balance to one's moral universe. For Katz, the imperative to mend one's image of how the moral universe ought to be overpowers rationality, especially the rationality that is thought to bind deterrence to the likelihood of punishment.[16]

Evidence of the contradiction can be found in critical analyses of forensics, but it seems that very few have articulated it directly. In Kruse's ethnographic account, a forensic technician is quoted who is obviously aware of the dilemma: "Those who are sober, the thieves that is, of course they know what they are doing, they protect themselves the whole time. So there's not a lot we can do, unfortunately. That's why the success rate [of solving burglaries with DNA and other forensic evidence] is so low. But the dopeheads and those who are a little high and a little drunk, they usually botch up."[17] At other points, the experts in Kruse's study note that "troublemakers" learn about forensic techniques and take steps to evade detection.[18] Gerlach also puts his finger on the problem in noting the limits of DNA technologies and, perhaps more important, how they cannot necessarily be accorded the magical powers so often attributed to them within the public imagination: "Criminals can and will develop tactics to evade the 'ultimate identifier.' Genetic justice will not end crime; however, it will change how crimes are carried out and detected."[19]

Bio-forensic technology thus finds itself locked within irresolvable contradictions: the criminalized behaviors that it purports to control (and, ideally, eradicate) are an indispensable component of its very assemblage. Without "crime" or the "criminal," bio-forensics is redundant. The criminal actor is reduced to being governed by rationality or irrationality, but neither construct of the subject coheres with the logic and practices of bio-forensics.

The Myth of the "Immaculate Database"

In their appraisal of biosocial criminology, Carrier and Walby criticize the perspective for thinking it possible to access pure, immaculate data and for assuming a one-way street that runs from such data to theoretical arguments.[20] In other words, biosocial criminology purports to be based on data—and readings of those data—that are unmediated by sociocultural forces. An analogous point could be made of identification databases, hence the notion of the "immaculate database." Bio-forensics appears to many as objective and machine-like in its workings. As such, it is presumed to operate in ways that are outside, or somehow beyond, sociopolitical relations.

Bio-evidence appears objective for various reasons. According to Gerlach, the objectivity of DNA (and presumably, forensic techniques with similar qualities) is produced by speaking in terms of "odds calculations"—or probabilities, likelihoods, chances, and so on—concerning matches.[21] Such terminology transforms the interpretation of DNA evidence into a mathematical operation, the epitome of pure abstraction. In exploring the history of fingerprint identification, Cole notes that fingerprint examiners actively constructed the fingerprint as "an empty signifier—a sign devoid of information about a body's race, ethnicity, heredity, character, or criminal propensity."[22] Fingerprint examiners distanced themselves from biological discourses of criminality because the latter were seen as locating the fingerprint within sociopolitical relations, thus tainting its appearance of objectivity. Given that the fingerprint often seems free of bias and thus "more factual," fingerprint examiners were quite successful in their constructivist endeavors.[23] The veneer of objectivity is achieved when the "signs" of bio-forensics—fingerprints, DNA samples, and so on—are reduced to objects that, found at crime scenes, do nothing but lead us to suspects and, possibly, perpetrators.

Bio-forensics, however, cannot but operate in social, political, cultural, and institutional contexts. The normalization of its use, its attachment to specific behaviors that have come to be regarded as "serious crimes," is a cultural phenomenon. Arguably, however, a more straightforward way to call the immaculateness of bio-evidence into question is by first noting that according its "signs" evidentiary value is contingent upon databases. The production of these databases, their contents, and access to them and their associated technologies, reveal the not-so-objective qualities of bio-evidence.

To borrow a theme from Bernard Harcourt's work on prediction, the chance that any particular individual will end up with a biometric identifier in a forensic database is not randomly distributed.[24] Rather, the likelihood of being recorded in a database is skewed by sociological factors, especially race, class, and gender. This skewed nature of biometric databases is not a new problem. Cole shows, for example, that one of the earliest uses of fingerprinting was coterminous with the colonization of India by the British. In distributing government pensions, British colonizers and administrators suspected that many people were impersonating deceased individuals in order to defraud the colonial state. Such frauds were, apparently, relatively easy to commit because British administrators had a hard time distinguishing one brown person from another. Fingerprints were thought to be an immutable imprint that revealed true identity and therefore capable of sorting fraudulent from legitimate claims. Fingerprint technology was thus incorporated as one strategy among others to administer the colonial state.[25] Needless to say, databases were populated with biometrics taken not from colonizers, but from the colonized.

Of course it was not long before fingerprinting technology found its way from civil administration into criminal justice systems. And because it is relatively quick and easy to take fingerprints, collection of them has spread to a wide range of minor offenses, such as fare evasion, public drunkenness, prostitution, and low-level property crimes. Insofar as those from minority groups are likely to be targeted by police for such offending behaviors, they are more likely to end up in forensic databases and be transformed into "recidivists." Cole expresses this well: "If arrest patterns are racist and classist, then the criminal identification databases themselves will contain disproportionate numbers of racial minorities, immigrants, and the poor."[26] Regarding disproportionality in databases, Lynch and colleagues note that on March 31, 2006, there were 3.7 million criminal justice profiles in the United Kingdom's

National DNA Database (NDNAD). According to calculations appearing in the *Guardian*, "37 percent of black men and 13 percent of Asian men in the nation are contained in the NDNAD, as compared with 9 percent of white men."[27] And in 2007 it was reported that 75 percent of Scotland's "young black male population will soon be on the DNA database."[28]

Disproportionality in the UK DNA database remains discernible. At the end of 2019, the number of samples from Black people in the database was 7.56 percent, whereas 2011 census figures (the most recent available) classified 3.01 percent of the population as Black. Conversely, the number of samples from white people was 77.76 percent, whereas 2011 census figures classified 87.17 percent of the population as white. To put this otherwise, while 25.36 percent of the nation's Black people appear in the DNA database, only 9 percent of white people are included. It is also worth noting that in the last count for 2019, the database contained 6,387,001 profiles.[29]

It is these databases against which crime scene samples are searched for possible matches, and their composition intimates what they can and cannot accomplish in practice: rather than dissolve crime, the main function of criminal databases is to identify those from particular social groups who recidivate.

The sociopolitically skewed nature of bio-forensics is also manifest in terms of access and costs. As various scholars have noted, forensic equipment and analysis amount to a set of tools that have been colonized by the state and prosecutors. Defense counsels and defendants may subject forensic evidence to cross-examination, but they are not often positioned such that they can orchestrate independent tests and expert testimony.[30]

Lynch and colleagues emphasize that the equipment to run DNA samples can be incredibly expensive, with some systems costing more than $200,000.[31] Moreover, because forensic laboratories are closely aligned with the states in which they operate, defense lawyers often have to rely on laboratories beyond the state that is prosecuting a case if they wish to independently test DNA samples. As Gerlach writes of the Canadian context: "[The] defence cannot use the same laboratory [that the state utilizes] and often must resort to expertise in another country, as happened in the Morin and Milgaard cases. However, independent laboratories charge between $800 to $3,000 per analysis, and expert witnesses charge around $1,200 per day for their testimony. For most criminal defendants, these costs are prohibitive."[32] Databases that are constructed and used to identify authors of criminalized acts are not immaculate; they do not magically fall from the sky or emerge independently of the social. Nor are they socially neutral, democratic tools. Instead, they are the product of a specific gaze, one that selects its field of vision along lines of race, class, gender, and so on. The contents of databases inevitably reflect political logics and institutional practices, which are indebted to legal and cultural frameworks concerning what constitutes a "crime problem," "serious crime," "criminality," and so on. Bio-forensic technologies include the socially excluded and exclude the socially included, and work to reproduce social exclusion rather than foster inclusion. Their portrayal and construction as immaculate—as if beyond the sociopolitical and capable of doing nothing but recovering past realities—conceals their politicized inputs and outputs. Rather than a resolution to crime, this is better framed as the manufacture and maintenance of a criminalized class, whose members are subject to biosurveillance and increased chances of incapacitation.

Rendering the Exercise of Individual Rights as Social Deviance

Our third critique revolves around the capacity of bio-forensics to produce a particular type of social deviance. More specifically, it transforms the protection of individual rights into a deviant practice, one that warrants suspicion of criminalized behavior or a lurking criminality. This is especially evident when one considers the use of "DNA dragnets."

A DNA dragnet is a costly (and often inefficient) investigative technique that starts with a DNA sample from a crime scene for which there is no match in a database.[33] Because there is no match, law enforcement officials embark upon finding one by asking hundreds, sometimes thousands, of individuals to voluntarily submit DNA samples. Individuals are likely to be selected if they "lived, worked, or were otherwise present in the general vicinity of the crime at the time it was committed, and they fit a basic description of the perpetrator."[34] Dragnets are most likely to be used when high-profile crimes—typically violent murders and sexual assaults—remain unsolved and investigators are struggling to turn up new evidence to progress a case.

Natalie Quan notes that DNA dragnets are problematic for a number of reasons. Because they target many individuals whom law enforcement officers have little reason to suspect, they are difficult to reconcile with rights that protect against unreasonable search and seizure.[35] Moreover, when combined with tools such as "DNA witness," they reproduce the idea that "race" is biologically given—or coded in genetic material. *DNA witness* is a technique that creates a visual image of a suspect from samples retrieved from a crime scene. It is purported that ancestry-informative markers (AIMs), which are genetic markers, can be used to project the "race" of a suspect.

AIMs are often used to identify and "cluster people into categories that roughly correspond to continents" and thus ancestral groups.[36] These ancestral groupings are then taken as indicative of "race." This ontologizes race—or construes a "race" as a discrete group prescribed by genetic difference from other groups; genetics-based ancestry research reveals "that all humans fall on the same uninterrupted gradient of genetic variation," but programs like DNA witness arbitrarily break this spectrum into discrete racial groups.[37] In other words, genetic research suggests an unbroken continuum of genetic variation; to impose lines or breaks upon this continuum is therefore an interpretive, speculative exercise. As Quan puts it, "whoever assesses the frequency of these genetic markers has made a subjective decision about, for example, how light skin must be in order for a person to be white."[38]

Gerlach also stresses that DNA dragnets can be taken as evidence of how bio-forensics threatens individual rights, especially the presumption of innocence, bodily integrity, freedom from self-incrimination, and freedom from unreasonable searches and seizures, because it accords priority to state interests, security, and technological crime management.[39] In a related vein, Lynch and colleagues note that as biotechnologies emerge and become normalized in criminal justice settings, "conceptions of criminal suspects and suspect rights" are recrafted.[40]

However, the problem is not only that developments in biotechnology accord priority to state authority at the expense of individual rights, or that they proffer a particular understanding of "justice" and what this looks like in practice. They go a little further in demanding that the honoring of, and attempts to enact, individual rights be understood as socially deviant and thus reasonable grounds for suspicion.

The Colin Pitchfork case is often recounted to illustrate the earliest use of DNA technologies and the DNA dragnet.[41] In 1983, the body of a fifteen-year old girl was found raped and strangled in a Leicester village. In 1986, the body of another fifteen-year old girl was discovered. DNA samples from both crime scenes were taken and found to be a match. The police initially suspected Richard Buckland, but DNA tests ruled him out. The police organized a DNA dragnet, taking samples from over four thousand men.[42] The dragnet did not directly identify the culprit, Colin Pitchfork. However, when an individual—apparently plagued by guilt—revealed to a small group of coworkers that he had provided a blood sample on Pitchfork's behalf, one of the coworkers notified police of the evasive move. A DNA sample was then taken from Pitchfork and was found to match the samples from both crime scenes.

Gerlach recalls three Canadian DNA dragnet cases, the first of which involved a serial rapist.[43] In talking about this dragnet, he notes that "during the initial investigation, no one refused to give blood. This was because of the obvious stigma attached to a refusal." The possibility of being stigmatized for refusing to participate is not unfounded; Gerlach cites a report in which a police official claims that if someone were known "not to give blood . . . he would be really, really unpopular." It is then noted that during a town meeting about the case, "two men spoke out against the mass testing as a violation of privacy; they were shouted down by the other residents, who supported the investigation." Finally, Gerlach notes that police warned that anyone refusing to participate in the dragnet "would face an intrusive background check."[44]

We have recalled these dragnet stories at some length because they reveal how the exercise of individual rights—freedom from self-incrimination, freedom from searches and seizures that lack probable cause, and so on—is stigmatized by bio-forensics. Bio-forensic investigation is heralded as an objective seeker and purveyor of the truth. Its mission is presumed to be of paramount importance, such that other ethical standards and principles can be dismissed, even rendered evidence of deviance if articulated or put into practice. For the police officer who warns of "intrusive background checks" into those who refuse to cooperate, individual rights appear as lacking substance, their exercise effectively construed as nothing more than "grounds for suspicion." In this view, there is something defective about an individual who expresses greater concern for rights than for the types of truth that bio-evidence and the state are invested in fabricating.

There is, however, more to the story. How the dragnet is put into discourse, debated, and socially received reveals that the stigmatization now appended to individual rights is deeply engrained within the sociocultural order and the psyche of its subjects. At the public meeting, the very few who raise the issue of rights to privacy are silenced; the police official does not appear concerned that those who do exercise their rights will be ostracized by the community, and if anything, the official recognizes the possible "excommunication" as an asset to the investigation. Regarding the subject and its interpellation into a social order in which individual rights are stigmatized, Pitchfork's coworker is racked with guilt, somewhat akin to Raskolnikov, the protagonist in Dostoevsky's *Crime and Punishment*, whose deed comes to light not because it is discerned by an external authority, but because it is forced to the surface by the subject's internal moral crisis.

The dragnet reveals how bio-forensic technology projects particular images of conformity and, consequently, new images of deviance, some of which may open the door to entanglement with criminal justice. In discussing media portrayals of criminal justice, Ray Surette draws a distinction between "due process" and "crime control" narratives.[45] In the former, criminal justice is portrayed as a system in which the guilt of an accused person needs to be proven against a backdrop of inviolable procedural rules. Such rules exist to ensure that the rights of citizens are protected and that the exercise of government authority is regulated. In contrast, "crime control" demands that criminals be identified and rapidly processed by the criminal justice system. Swift punishment is advocated, and any concern with individual rights is associated with protecting and favoring the interests of "criminals." DNA dragnets take for granted the "crime control" narrative and reproduce it. They articulate the view that individual rights are a refuge for "criminality" rather than universal principles that curb the excessive power of states. Within the crime control narrative, recognizing rights then becomes tantamount to siding with "criminality"—a callous disregard for the social body—and that is enough to render one deviant, untrustworthy, and a risky prospect in need of surveillance.

SHADES OF (UN)CERTAINTY

Insofar as it comes to be perceived as infallible, bio-evidence can more readily claim itself capable of resolving crime. And to be sure, it has been quite successful in crafting an image of itself as unassailable. This is especially the case with DNA typing, but an aura of infallibility long surrounded (and continues to surround, albeit to a lesser extent) fingerprint matching and other criminalistic techniques—even lie detector testing. Popular shows like *CSI*, *Dexter*, and *Bones* all impute infallibility to forensic evidence. As Jay Aronson and Simon Cole point out, however, infallibility is also imputed to DNA typing from quarters that one might not anticipate, such as those who seek to reveal wrongful convictions via DNA evidence and thus set the innocent free.[46] Within narratives of forensic unassailability, it is often asserted that developments

FIGURE 3. The "lie-detector" or polygraph test. Polygraph examination is not sufficiently robust to pass the Daubert test and hence is inadmissible as evidence. Nonetheless, much of the public arguably perceives it as capable of discerning guilt and innocence. As with other crime science technologies, such as cranial measurement and brain scans, the faith that criminality can be detected surfaces. *Sources:* Kat Eschner, "Lie Detectors Don't Work as Advertised and They Never Did," *Smithsonian Magazine*, February 2, 2017 (image is from Wikimedia commons); and Christopher Bergland, "One Simple Way You Can Become a Human Lie Detector," *The Athlete's Way* (blog), *Psychology Today*, October 31, 2015 (image is from Ed Westcott/public domain).

in scientific techniques will correct problems within criminal justice: if absolute truth can be ascertained, wrongful convictions will disappear, corruption or the pursuit of subjective interests among officials will be detected (thereby realizing Beccaria's dream), and criminals will not be set free for lack of credible evidence.[47]

Science and technology scholars have explored how fingerprint matching and DNA typing is practically carried out and, through such analyses, have crafted narratives in which the possibilities for error are emphasized. Much of this scholarship owes a debt to Bruno Latour's notion of "black boxing," which refers to construing technologies as infallible not on the basis of how they actually work, but by concealing their inner workings and processes from scrutiny.[48] Concealing the inner workings of any given technology occurs in various ways: institutional processes may make it impossible for outsiders to observe how a technology works, rhetorical assurances that any problems with the science have been worked out may be provided, or protocols may be designed to assure outsiders that rules and procedures are in place to prevent error.[49]

In opening up the "black box" of bio-forensics, many of the arguments advanced by science and technology scholars are quite persuasive. They have shown, for example, that fingerprint analysis is an interpretive exercise rather than a robust scientific method and thus open to costly errors such as false conviction. Similarly, DNA-related techniques and analysis are not as clear and straightforward as they are often purported to be; samples need to be collected and transported in particular ways, it is possible for samples to become contaminated during transit or once they enter laboratories, analytical techniques are messy in practice, outputs require interpretation, and it is not always possible to rerun an analysis for the sake of verification.[50]

There is much that is commendable here, and science and technology scholarship has been repaid kindly for investing in Latour's "black box" (and its related concepts). Nevertheless, drawing attention to processes that transpire behind the scenes entails sacrificing some theoretical interpretations and possibilities. Given that DNA typing and discoveries of faulty evidence leading to erroneous convictions are often coterminous with high-profile crimes, it is not surprising to find that science and technology scholarship regularly operates with a hegemonic conception of "serious crime," discernible in the heavy focus on cases involving murder and violent sexual assault. Everyday, lower-level criminalized behaviors—or the types of offending that tend to saturate the criminal justice system—do not make much of an appearance because they do not typically involve forensic evidence.

At least two consequences follow when dominant constructions of "serious crime" are accorded priority. First, it would seem that the notion of "injustice" comes to be understood in a fairly circumscribed manner, especially insofar as it is conflated with "erroneous conviction" (for a "serious" criminal offense). In Cole's account, a "false conviction" amounts to the "worst nightmare" of bio-forensics and criminal justice: "[The] fingerprint examiners' worst nightmare, an erroneous identification—resulting in the criminal justice systems' worst nightmare, a false conviction."[51] The skeptic in us doubts that this is the case; it seems that criminal justice can sleep quite well at night despite false convictions. To be sure, the *discovery* of false convictions may cause night terrors, especially if those convictions become highly publicized and politicized. Moreover, allowing "false convictions" to colonize the notion of "injustice" renders invisible much of the injustice that is endemic to the ordinary operation of criminal justice: its overwhelming focus on those from marginalized social locations (both discursively and in practice), all those moments in which "counter-law" (e.g., discretionary uses of authority) trumps formal protections, and the relative powerlessness of the powerless to defend against the encroachments of the state. Of course, it is reasonable to assume that science and technology scholars are well aware of these issues, and it is reasonable to assert that false convictions are fundamentally unjust. A problem emerges, however, insofar as injustice is reduced to false conviction, a discursive move that presupposes the overarching legitimacy of the state and its criminal justice system.

Second, and arguably a more disconcerting problem, there is a tendency to make inferences about broad trends in criminal justice upon analyses of how bio-forensic technology and "serious crime" interface. Gerlach, for example, expresses concern that criminal justice systems may fall prey to scientization, thereby coming to operate as "genetic justice systems":

> DNA analysis and other developments in forensic science are not simply another element added to the existing legal process; they are changing the very nature of the legal process by redefining the nature of the game and transforming the necessary habitus. . . . Within the scientized field of the courtroom, the defence lawyer is at a disadvantage. His or her main enemy is the alliance between forensic laboratories and prosecutors—an alliance that allows prosecutors to push for higher levels of scientific discourse within the courtroom and draw from the charisma of genetic science.[52]

In addition, "[It] is no longer an issue of scientists replacing lawyers and judges as arbiters of justice; rather, it is a matter of everyone getting involved in adopting a scientific discourse and scientific standards of evidence and proof."[53] In a similar fashion, Lynch and colleagues have argued that DNA-related technologies run the risk of trumping all other forms of evidence, thereby emerging as the ultimate "truth machine." As a truth machine, DNA effectively determines how other forms of evidence are to be interpreted. It is now possible, for instance, to consider whether fingerprint evidence is consistent with DNA evidence. Moreover, it is DNA that often stands as the final arbiter when a case involves disparate accounts from witnesses, or those who offer testimony in the courtroom.[54] For science and technology scholars, the irony of DNA evidence standing as the ultimate yardstick against which all other evidence can be assessed stems from a core axiom within the philosophy of science: what seems true today may not be true tomorrow; one never arrives at a point of absolute certainty via science.

The problem with claims that DNA technologies threaten to scientize criminal justice, or that DNA evidence emerges as the adjudicator of competing truth claims, begins with their limited applicability. To be sure, many science and technology scholars acknowledge that most criminal justice cases do not involve DNA evidence: "In the early years around 80 percent of cases involving DNA were sexual assault cases, and this is still the principle application of forensic DNA typing."[55] "Despite efforts to expand its use, DNA evidence continues to be collected in a relatively small proportion of criminal cases. And . . . even when DNA evidence is available, its value and significance depend upon other evidence."[56] And yet there is a tendency to imply that cases involving DNA can be read as portending general trends in criminal justice. We argue that there are effects that follow from criminal justice systems incorporating DNA evidence, but these are quite distinct from those enumerated earlier. If criminal justice systems were "scientized," and if the certainty associated with DNA operated as the ultimate truth machine, far fewer convictions would transpire. Further, we would see fewer arrests proceeding to prosecution, more cases contested and hence more trials, and a greater proportion of not guilty findings in cases that do go to trial.

Somewhat paradoxically, the issue is that DNA evidence and the aura of infallibility by which it is surrounded do not extend throughout the criminal justice system. From the perspective of those invested in making arrests, securing convictions, and sending people to prison, an excessive dependence on DNA evidence would impede the everyday workings of criminal justice. This is not to suggest that DNA and other types of bio-evidence can establish "truth," or that they should be extended such that they saturate the criminal justice system. Rather, it is to explore how bio-forensic technology plays a role in actively constructing thresholds of truth and shaping how and when they ought to operate within criminal justice.

Imagine, for example, an instance in which an officer finds drugs in an individual's car during a highway stop. Imagine further that the motorist later claims the officer planted the prohibited substance. In the courtroom, we end up with two competing narratives: the officer's, who claims it was in the motorist's car, and the motorist's, who claims the officer put it there. There is no DNA or forensic "truth machine" in this scenario, but we somehow doubt that a judge and/or jury will privilege the motorist's narrative over that of the police officer.

And in a not-so-hypothetical example, between 2013 and 2016, one of us was conducting an ethnographic study of New Zealand's district courts (which are lower-level trial courts that adjudicate the vast majority of criminal cases). As part of that study, much time was spent unobtrusively observing hearings and sentencing decisions. One such hearing stands out in this context. According to the "facts of the case," the male defendant entered a woman's clothing store and asked to try on some items. The store attendant, a young woman, assisted the man. After a while, she noticed that the man was still in the changing room, so she peeked under the changing room door to see if everything was OK. She noticed that the man had his pants around his ankles, and that some pornographic magazines were laid out on the floor. Not quite sure how to proceed, she called the police. When they arrived, the arresting officer peered over the top of the changing room door and saw the man with his penis out. He forcefully opened the door and arrested the man.

During the court case, the officer testified that he saw semen on the man's hand. The defense questioned if the alleged semen had been subject to any forensic testing. The defense raised this issue because lubricant packaging was found in the dressing room, and the defendant claimed it was lubricant that was on his hands, not semen (as he was only preparing to masturbate). The police officer was adamant that it was semen because he had fathered three boys and so knew semen when he saw it. This was satisfactory to the court, which found in favor of the prosecution. The defendant was set a sentencing date, and presentence reports were ordered.

Our point in recalling this case, as anecdotal as it may well be, is to suggest that everyday criminal justice is not necessarily invested in framing or utilizing bio-forensic analysis as the ultimate "truth machine." In this case, it is the officer's three sons (and such fatherhood status somehow translating into unassailable evidence concerning when a substance is semen) that trumps forensics. That the defense asks for a forensic comparison could certainly be taken as verification of Gerlach's "scientization" thesis, a shift in the legal habitus. However, the other major players in the case—prosecutor, testifying officer, and judge—do not appear to have been structured by this same habitus.

In our view, the effect of bio-forensics in criminal justice contexts has not been the *general* imposition of a new, ultimate truth machine, but the crafting of different, if not incommensurate, notions of "beyond reasonable doubt." Bio-forensic technology tells us when higher levels of certainty can be expected and, by implication, when we can dispense with the types of certainty it promises.

To put this another way, DNA and other types of bio-evidence may operate as a "truth machine" in particular areas of criminal justice, but if this logic were extended, criminal justice would stumble. In order to avoid stumbling, various thresholds of truth have come to operate across and within criminal justice. We suggest that the rise of DNA evidence plays an important role in constructing the category of "serious crime" and effectively appends specific standards of truth to offending types in which DNA evidence is likely to exist. Conversely, cases in which DNA evidence is noticeable by its absence are—by virtue of this absence—construed as "not so serious" and can thus be calibrated to "nonscientific" standards of truth.

Evidence for this view of multiple, possibly incommensurate interpretations of "beyond reasonable doubt" or "legal truth" is discernible in critical accounts. In Kruse's study, one of the judges interviewed more or less "confesses" to the problem we have described:

> Society, he explained, "can accept a fine being imposed based on a false confession," but it cannot accept that someone would be given a long prison sentence for a crime they did not commit. Thus, the more severe the crime for which a defendant is being tried, the higher the standards on the investigation. When someone falsely confesses to shoplifting and is fined, it is of minor consequence to them and a small risk to society. But a confession in a murder case cannot be taken at face value, as miscarriages of justice in murder cases have much more severe repercussions.[57]

Underpinning this view is a continuum of "legal truth": at one end, it is necessary to corroborate accounts with some kind of truth machine, so much so that even the confessions of defendants cannot be taken at face-value, but at the other end, one finds the possibility of false confessions somehow surpassing legal requirements of truth. All of this would appear to hinge on a hierarchical ordering of crimes according to "seriousness." Those crimes designated as "serious" demand relative certainty, whereas false confessions can be tolerated provided they lead to convictions and punishment for crimes deemed less serious. Bio-forensics corresponds to the former category but is rendered increasingly unnecessary as one descends into zones of lesser seriousness.

These different thresholds of truth, which correspond to "seriousness of offending" and possibilities of bio-evidence, return us to the concept of injustice (or justice). Injustice is effectively designated a possibility within a relatively small area of the criminal justice system—namely, that part that sees murder cases, some "sexed" crimes, and perhaps a portion of property offenses. Although admitted as a possibility in "serious cases," there is a presumption that injustice cannot eventuate in those areas that have been designated as low stakes. Or if injustice does transpire in the mundane regions of criminal justice, it can be dismissed as relatively inconsequential and thus tolerable.

As do other branches of crime science, bio-forensic technology sanctions a cavalier attitude toward those whose freedoms and rights may be at stake. This attitude is especially encouraged in the realm of everyday, "minor crimes"—which just so happen to constitute the vast majority of cases processed by criminal courts. As the judge's remarks and sentiments reveal, these everyday punishments can be fundamentally unjust and yet safely excused because their social and individual consequences are purportedly of minor consequence. We would like to hear from those on the receiving end of punishments that follow from false confessions deemed unworthy of further (bio-forensic) verification. We somehow doubt that they will evaluate injustice in the same way as those who work the machinery of everyday criminal justice. It is all too easy to minimize the impact of injustices meted out by criminal justice when you are unlikely to find yourself on the receiving end.

CRIME, VICTIMIZATION, EQUIVOCATION: THE POLITICS IN THE DETAIL

In the remainder of this chapter, we switch tracks to "biosocial criminology." Those who advocate for this variant of criminology routinely construe it as more than a simple revival of Lombroso-style biological criminology, sometimes going so far as to claim it represents a "revolution" in criminological thought. The perspective has been heavily criticized by radical constructivist criminology, especially by Carrier and Walby, who argue that it amounts to little more than a repackaging of Lombroso. In their reading, biological and biosocial criminology belong to the same underlying paradigm; at best, the latter can be said to play around with a few of the former's minor premises.[58]

We agree with this line of critique but wish to extend it by focusing on the equivocations of biosocial criminology. It is these equivocal moments that, we argue, profoundly reveal the politics by which biosocial criminology is animated. We use equivocation in its general philosophical sense, wherein it refers to appending multiple meanings to the same term or underlying construct that is facilitating analysis. Biosocial criminology does this with most of its major categories. In this section, our focus is on the notions of "crime" and "victimization." Following this, we examine how biosocial criminology variously frames the relationship between science and society, the sanctity of numbers, and the notion of "adaptation."

The equivocations within biosocial criminology are especially evident in Anthony Walsh's *Race and Crime*, an account that seeks to popularize the allegedly new approach by providing an overview of its major lines of argumentation.[59] Given its power to illuminate the illogic of biosocial criminology, this text provides a thread by which our critique is organized.

In *Race and Crime*, Walsh begins by attempting to show that "race" is not a social construct but a concrete reality that correlates with crime. For him, the reality of race can be discerned in the work of population geneticists, who claim that genetic markers can be used to map ancestry and reveal the existence of various ancestry groups.[60] Walsh takes this as evidence that refutes the constructivist position on race: because people can be grouped according to common ancestry, race cannot be discursive. The supposed nondiscursivity of race is a leitmotif of biosocial criminology.[61] Given the recurrent effort to frame race as an ontologically given reality, it is worth recalling the problem that Quan identifies in treating ancestry maps as evidence of the reality of "race": to carve an unbroken *continuum* of genetic variation—which is what ancestry research posits—into racial groups is to construct, and impose, racial classifications that are not given by ancestry.

However, not only is ancestry research misread. Walsh and other biosocial criminologists simply fail to understand the logic of constructivism, often confusing it for a colloquial notion of "semantics"—that is, "different words for the same thing."[62] Even if one were to accept that a concrete referent can be appended to "race," it does not follow that race can be reduced to this, as if words were nothing but arbitrary linguistic mirrors for some fragment of "reality." For many constructivists, the point is that the category of "race" (and language, "scientific discourses," etc.) cannot but operate as a cultural force that mediates practice. As Carrier and Walby suggest, ignorance on this point goes a long way toward explaining why "biosocial criminologists are often accused of racism."[63]

Upon this faulty reading of constructivism, Walsh proceeds to link "race" and "crime" by showing how supposed racial differences correlate with differential involvement in crime. Utilizing official data from the United States, such as the Uniform Crime Reports, Walsh posits that the crime rates of "African Americans" are greater than those of whites, and that the crime rates of whites are greater than those of "Asians."[64] Much of the text then focuses on accounting for the crime rates of Black Americans via the logic of biosocial criminology, which is said to be constituted by three broad orientations: genetics, evolutionary psychology, and neuroscience.[65] All of these orientations, it is claimed, assume that biology and environment interact to produce behavioral outcomes, some of which are unproblematically classified as "criminal."[66]

As one might anticipate, a range of statistics, studies, concepts, correlations between variables, and so on is marshaled in support of biosocial criminology. However, it does not take long to realize that Walsh's text is premised upon a series of conceptual and logical inconsistencies that fundamentally undermine its specific arguments and details. The categories that are central to the architecture of biosocial criminology are given various meanings throughout, and it can hardly be said that such shifts in meaning follow from some "external reality"—despite the repeated refrain of biosocial criminology that it is dragged to its arguments by a reality that is held to be "objective," or unmediated.[67] The equivocations, however, become much more comprehensible if it is accepted that biosocial criminology amounts to a political project, one that is overwhelmingly preoccupied with inscribing criminality on particular bodies.

In Walsh's initial definition of crime, victimization appears as a corollary notion: "The criminal acts that most concern us are universally condemned and are inherently evil (*mala in se*), the litmus test for which is that no one wants to be victimized by them. *Mala in se* crimes such as murder, rape, and theft of resources, are crimes that militate against the evolutionary imperatives to survive and reproduce."[68] Biosocial criminology, then, is concerned with crimes that are thought to be "inherently evil" rather than merely prohibited by statute (*mala prohibita*). Importantly, Walsh posits that there is a litmus test for identifying those crimes that are evil in themselves, which involves not wanting to experience them. Moreover, crime and victimization can be assessed in reproductive terms: acts that threaten one's will or ability to survive and reproduce are profoundly evil; therefore, to experience them is to be victimized.

In accounting for the supposedly higher crime rate among Black Americans in the United States, Walsh attributes this to slavery and racism: "I address racism first because of my belief that the cultural roots of the horrific rates of antisocial behavior in the inner cities [i.e., Black American communities] lie in slavery, the racism slavery generated, and the oppositional subculture that arose in the African American community in response to it."[69] Throughout the text, Walsh repeatedly comes back to the problems of slavery, racism, and the discrimination that whites have practiced against marginalized social groups in the United States. Slavery is described as "the most odious practice ever instituted by human beings"; discrimination is construed as part of a "history of white malfeasance" and "horrific."[70] It is acknowledged that "American Indians" were driven from their lands and that their population numbers have been "severely denuded by wars, the introduction of 'European' illnesses, and general mistreatment."[71]

Such descriptions of slavery, racism, and discrimination would seem to be entirely consistent with Walsh's initial definition of "crime," and the atrocities committed by whites that Walsh describes clearly pass his "litmus test" for what he considers *mala in se* crimes. And yet never is this violent history of white America directly framed as "criminal." Suddenly, behaviors that are criminal by his own standards are understood as "odious" and "horrific." Walsh effectively describes the genocide of Native Americans, which would appear to "militate against the evolutionary imperatives to survive and reproduce," but such criminal activity is rendered as "mistreatment."[72]

In discursively relocating the crimes and atrocities committed by white people to a range of alternative signifiers, Walsh removes any need to subject whites to the kind of biosocial theorization that he wishes to utilize in relation to Black Americans. At no point, for example, does Walsh seek to develop a biosocial account of colonization, slavery, or discrimination. This selectivity concerning to whom biosocial theory ought to apply is not peculiar to Walsh. In the early 1990s, when talk of a "warrior gene" surfaced—based on the work of Han Brunner and colleagues, who claimed that deficiencies in MAO-A were responsible for criminal behavior—and was used to explain the supposed propensity among indigenous populations to engage in violent behavior, critics were quick to respond that biosocial criminologists appeared strangely uninterested in identifying genes that would explain the colonialist desire to subjugate indigenous populations and engage in genocide.[73]

In aligning the category of crime more or less exclusively with "African Americans"—and simultaneously severing it from whites—Walsh creates the space needed to append multiple meanings to the concept of victimization. Rather than figuring as victims of slavery and racism,

Black Americans are discursively positioned as suffering from a "cult of victimology." This is an aspect of the "oppositional subculture that arose in the African American community" due to racism, and which initially served protective functions but is now held to be dysfunctional and "maladaptive."[74]

Walsh is aware that every minority group in the United States and elsewhere has experienced discrimination. Native Americans were nearly eradicated by violent colonization; Asians, especially "the Chinese, have suffered a great deal of prejudice and discrimination in the United States"; and "Jews are well aware of their ancestor[s'] historical victimization."[75]

However, these other groups—Asians and Jews—have realized something important that apparently eludes Black people—namely, that these forms of victimization are merely historical. As Walsh expresses the point, Jewish people "get on with their lives in whatever country they find themselves, realizing that to dwell on the past is to sabotage the future."[76] When referring to socially marginalized or minority groups, it is the inability to subjectively (and collectively) transcend past experiences in which one has been victimized by crime that amounts to victimization. In other words, victimization no longer entails the suffering associated with a criminal event (or events), but instead is the tendency to dwell on any such experiences. Initially understood as a social relation, victimization is recast as a psychological problem. Once framed in psychological terms, victimization is understood as an "attitude," an erroneous "self-perception": "Many blacks have succumbed to a 'cult of victimology.' . . . [This] tendency to view themselves as victims sabotages the ethic of self-reliance among African Americans, and grants them permission to condone weakness and failure."[77] Black Americans, to put it colloquially, need to "get over it": "The most important characteristic that applies mainly to blacks involves the transcendence of 'me as victim' attitudes. . . . If blacks can dump the blame game, they will have gone a long way towards changing their lives."[78]

It is rather obvious that these two understandings of victimization cannot proceed from some "unmediated reality," but rather derive from the political standpoint that Walsh occupies. In its initial articulation, victimization is linked to crimes that interfere with reproductive capacities. When it must be conceded that racialized groups have experienced the most atrocious forms of victimization according to that very definition, Walsh produces a second meaning for the term. In this second iteration, victimization is construed as "historical," "subjective," a "delusional belief" (evidenced by the notion of "cult"), a "false perception of self."

The definition magically changes depending on which social groups are the subject of discussion!

There is, of course, another major dilemma that equivocating on the concept of victimization entails. If victimization is something that can and needs to be "transcended," then why bother developing a biosocial criminological account of "crime" at all? According to Walsh's logic, "crime" is always "in the past"—it cannot but be a historical event. Given this, why are not all victims who "dwell on," or experience suffering due to experiences of victimization, not regarded as suffering from a "cult of victimhood" and simply advised to "get on with their lives"?

As several scholars have noted, biological and/or biosocial criminology routinely draws an absolute distinction between "criminal" and "victim," "them" and "us," the "unworthy" and the "worthy."[79] We posit that Walsh needs to distance Black Americans (and other marginalized social groups) from the "victim" category because, in the absence of doing so, any simple distinction between "criminals" and "victims" collapses. White oppression would have to appear as "criminal," and those who are overrepresented in criminal justice figures would have to be cast as "victims." Acknowledging such complexity, however, threatens the fundamental narrative and interventionist logics that biological crime science pursues. Far from being dictated by "reality," biosocial criminology is a political discourse that works to normalize the power of dominant groups to control the bodies of those who belong to racially marginalized groups.

Walsh claims to be "more of a 'pull your socks up' conservative than a 'stop blaming the victim' liberal."[80] Evidently, however, this logic only applies to those from social groups that have been forced to endure the most "odious" and horrendous forms of racism and discriminatory practice. Conversely, such conservative logic is not to be applied to, say, those from privileged social groups who experience "criminal events" as "victims."

OTHER CRIMINAL EQUIVOCATIONS: SCIENCE AND SOCIETY, NUMBERS, ADAPTATION

There are at least three other equivocations in the discourse of biosocial criminology that are worth drawing attention to. The first concerns how the relationship between science and society is conceptualized, the second revolves around the use of numbers, and the third involves the notion of "adaptation."

"Science and Society" versus "Science/Society"

Biosocial criminology routinely operates with two conceptions of how science and society are related. In the first conceptualization, science is seen as independent of society. It is conceived of as a "pure" space, a world unto its own in which no speech acts are, or can be, prohibited. As a quarantined zone of discourses and statements, science must "seek the truth, and nothing else."[81] As it is about "truth"—even truths that may be "unpleasant"—science can never be held accountable for being "misused" within the social realm, and it can never be taken to task for its political implications, no matter how racist, sexist, classist, or ableist they may be.[82] If anything does go awry in the social, science is presumed innocent because of its empirical and logical "purity."

This interpretation of the science-society relationship typically appears in the opening sections of biosocial criminology texts, and it often operates as a self-given license for authors to articulate a range of ideas that could hardly be described as neutral or apolitical. For example, Randy Thornhill and Craig Palmer's "evolutionary" explanation of rape—in which it is posited that rape is consistent with natural selection and its prescribing of behaviors that promote the continuation of genetic lines—illustrates well the tendency to seek sanctuary in some imagined realm of nonpolitical, "pure science."[83]

In Walsh's account, the idea of a space that can only be inhabited by pure science is necessary to accord "race" an ontological status and to divorce biological models from their nefarious eugenic history: "The horrors of Nazism are endlessly evoked as examples of the dangers of biological theories, but the nightmare of racial purity and ethnic cleansing that have bedevilled us throughout human history did not wait for Gregor Mendel or Charles Darwin to sanctify them. Nazi theories of racial superiority did not rest on any kind of reputable science, but rather on a quasi-mystical nationalism that hypnotized the German people."[84]

Nicole Rafter has suggested that Nazism constitutes "criminology's darkest hour," and reveals the numerous links between biological criminology and the racialized, eugenic practices of the Nazis. The Nazi party, for example, nationalized local criminal justice programs that collected medical and anthropometric data on prisoners, their families, and their associates, and used such information to identify "habitual criminals" or "incorrigibles."[85] Many people so identified were murdered by the Nazis. Moreover, criminal justice officials and scientists collaborated with the Nazi regime. As Rafter notes, the Third Reich was "far from hostile to science; rather, at every step of the way it called on science for verification and legitimation—just as science called on it, and for the same reasons."[86]

If one acknowledges this version of the historical record, it becomes difficult to accept biosocial criminology's position that science and society can be regarded as separate entities, and that any ethically problematic practices that appear to follow from science can be attributed to "non-reputable" science.[87]

In any case, it is not our intention to get bogged down in a debate over whether science is rightly framed as a discrete entity and therefore "innocent" if or when appropriated to support social practices that many might find morally questionable. More important in the present context is the tendency within biosocial criminology to embrace an interventionist logic the moment it thinks it can orchestrate practices that are presumed to benefit the social.

In the work of Michael Rocque and colleagues, who also promote biosocial criminology, it is conceded that biological theories were used to sanction eugenic policies, but then asserted that any such tendencies can safely be regarded as historical. This is taken as a guarantee that biosocial criminology currently operates—and will continue to do so in the future—as a benevolent social force: "We attempt to show that this new biological crime prevention is vastly different from the biological strategies of the past. Far from advocating unethical eugenic measures, this work focuses on improving the environment to promote healthy biological development in early life."[88]

In their response to Carrier and Walby's radical constructivist critique of biosocial criminology, Anthony Walsh and John Wright are adamant that science—presumably the "reputable" kind, which they are somehow able to distinguish from not-so-reputable forms—is capable only of intervening in social life in ways that are unequivocally good: "Instead of discovery, correction, and counter-correction, which the process of science offers, postmodernists take a position that denies any objective reality—a position that inadvertently leaves powerless individuals to fend for themselves against disease as well as criminal victimization. In this sense, postmodernists simply overlook the quantifiable improvements in human life brought about by science, including research on how best to intervene with at-risk youth and dangerous adults. Postmodernism offers nothing comparable to the contributions empirical science has made to the quality of life now experienced in many countries."[89] In this second conceptualization of the science-society relationship, science is reduced to a presumed capacity to address social problems. The boundary between science and society is now dissolved, with science intervening in ways that could never generate injustice, produce morally problematic outcomes, or serve particular political interests. And it is only this benevolent range of possibilities that the biosocial criminologist claims or acknowledges.

One could easily demonstrate the ridiculous simplicity and absurdity of this second conceptualization of science and its relationship to society: What should we make, for example, of all the scientific advances that have gone into the Industrial Revolution, now known to be destroying the planet? Or into prescription drugs that are later discovered to cause major health problems?

Somewhat ironically, Walsh and Wright's simplistic reading of science is betrayed by their own reference to "counter-correction": if science amounts to an unambiguous social good, then why would it ever need to "counter-correct" its initial "corrections"?

Arguably, however, the problem is not so much that biosocial criminology fails to recognize the dialectic, ambiguous nature of science—the tendency for its benefits to be accompanied by numerous, and often hidden, costs—but that its oscillation between these two views of science-society amounts to holding a position that is simply untenable, one that cannot be justified on "empirical" or logical grounds. Science is either "pure" or reduced to concern for the "truth," in which case it ought not to be presumed capable of addressing political and social matters (as this involves "interpretation"), or it is replete with sociopolitical significance, and therefore its statements can never be exempt from scrutiny for their political implications. The biosocial criminologist shifts between these contradictory positions according to political objectives and need: science is reduced to objectively derived, "pure" statements when race is to be ontologized, racism and discrimination exonerated, men's rape of women naturalized, and so on; it is construed as sociopolitical when biosocial criminologists think they have identified a policy or practice that seems promising.[90]

The two conceptualizations of science and society, then, make sense when recognized as belonging to a broader rhetorical strategy that seeks to strip scientific discourses of their normative ambiguity, the indeterminateness of their sociopolitical implications and effects. Biosocial criminology needs to assert that the "science" it espouses is somehow independent of social relations—but simultaneously a benevolent social force—precisely because it harbors very obvious rationales for policies and practices that are racist, eugenicist, and genocidal. Its oscillation on the difficult, science-society problem is a political disclaimer of sorts: if horror eventuates, this is not due to the qualities of our product, but a consequence of irresponsible consumers who did not read the (ingenuine, contradictory, and in very small print) warning label.

Playing the Numbers Games

Biosocial criminology is in awe of the "more fundamental sciences," especially biology and its subdivisions.[91] If left to its "sociological" devices, the biosocial criminologist asserts, criminology will forever remain a hopeless enterprise, incapable of cracking open the "crime problem." Given this "science envy," it is not surprising to find a heavy dependence on statistics and numbers within biosocial criminology.

Numbers are not necessarily problematic by default. Within biosocial criminology, however, the incorporation of numbers is often underpinned by different assumptions concerning their meaning or ability to access the real without mediation. And the way in which numbers are conceived shifts according to what they are being used to establish.

As noted previously, Walsh and other biosocial criminologists are adamant that there is a reality that corresponds to the concept of race, and that some racial groups are more criminal than others. Black Americans, it is held, are more criminogenic than whites, and whites are more criminally inclined than Asians.[92] To support these positions, official figures are often reproduced. Here is Walsh, for example, demonstrating the disproportionate volume of crime committed by Black Americans:

> According to the 2000 Census, African Americans constitute 12.8 percent of the U.S. population. . . . However, according to 2002 Uniform Crime Reports (UCR), the percentage of blacks arrested for each Part I index crime in 2001 were: murder (48.8%), rape (34.3%), Robbery (52.5%), aggravated assault (33.3%), burglary (29.9%), larceny/theft (32.6%), motor vehicle theft (39.3%), and arson (24.9%). . . .
>
> By way of contrast, whites are underrepresented relative to their proportion of the population. Whites constituted 82.2 percent of the American population in 1999 . . . , with arrest rates for murder (48.9%), rape (63.1%), robbery (46.0%), aggravated assault (64.4%), burglary (68.1%), larceny/theft (64.9%), motor vehicle theft (58.0%), and arson (72.4%).[93]

These figures are provided without caveats, cautionary notes, or acknowledgment of limitations. That these are official figures that will, at least to some degree, reflect institutional biases and processes, is never mentioned.[94] None of the crime categories is seen as in need of further definition. That these particular crime types have generally come to be understood as constituting "serious crime"—to the exclusion of, for example, white-collar crimes or state crimes—does not appear to warrant any comment.

In a later chapter on "extraordinary" crimes, however, Walsh is forced to confront figures on "white collar crime," in which whites would appear to be overrepresented. It is acknowledged that white-collar crimes can be devastating in their effects: "The S & L scandal [in the 1980s] cost the U.S. taxpayer more that [*sic*] $470 *billion*, which is more than all the conventional bank robberies in U.S. history, and the Ford Pinto case cost the lives of over 700 people, more than all the victims of serial murderers combined in the 1970's."[95] And, it is acknowledged that such crimes reveal a "different racial story": "An analysis of 1,094 white-collar criminals processed through the federal courts in the 1980s found that the demographic characteristics of the typical high-status white-collar offender differed radically from those of street criminals. Antitrust offenders were 99.1 percent white, and securities fraud offenders were 99.6 percent white."[96]

But unlike the Uniform Crime Reports, these figures concerning whites and white-collar crime, according to Walsh, need to be read with caution. One of the problems with such figures is that they hinge upon specific definitions of white-collar crime—definitions that only include acts of "high status individuals acting on behalf of his or her corporation."[97] This excludes things like fraud, counterfeiting, employee theft, and so on. Moreover, we also need to consider the role of "opportunity" in relation to these figures: "Anyone is eligible to commit a street crime or lower-level white-collar crime, but only a very few are eligible to commit upper-level corporate crime."[98]

We have quoted Walsh at some length not because we accept that there is an appropriate way to access reality through numbers, but to show the disparate assumptions that are made concerning the use of numbers. When the biosocial criminologist wishes to establish a relationship between "race" and "crime"—that is, to posit that Black Americans are more criminally inclined than other groups—official figures are wheeled out minus any critical sensibility. Those numbers, it is assumed, speak for themselves and can therefore be taken as an unmediated representation of reality. Conversely, when figures demonstrate the concentration of whites in particular offending types—thereby revealing a "different racial story"—a critical consciousness mysteriously moves to the foreground.

All of a sudden, how things are defined matters—and this notwithstanding the biosocial criminological refrains that constructivism has nothing to offer, amounts to mere semantics, and so on. The category of white-collar crime needs to be expanded to ensure it is not restricted to upper-class whites, and factors such as "opportunity" need to be factored into the account. That the costliest forms of white-collar crime are committed almost exclusively by whites, and what this "opportunity structure" might say about the intersection of inequalities grounded in class and race, does not appear to interest Walsh. As far as biosocial criminology is concerned, if it were found that 99.1 to 99.6 percent of corporate elites (or those in a position to commit major antitrust and/or securities fraud violations) were white, then the problem of upper-level white-collar crime would vanish because the demographics and offending ratios within the corporate world would be synchronized.

Approaching numbers with two sensibilities—uncritical acceptance of figures that suggest Black criminality versus critical appraisal of figures that reveal criminality among elite whites—cannot be logically justified. And there is no logical reason to assume that crime is problematic only when disproportionately concentrated among specific groups.[99] Problems associated with officially produced crime statistics—their inevitable dependence on institutional biases, definitions, practices, and so on—are far too obvious to ignore. And why powerful, white elites from the corporate world are almost exclusively responsible for committing the costliest forms of white-collar crime is also far too obvious to ignore. That Black Americans—along with other social groups, to be sure—are discriminated against and thus excluded from the upper echelons of the corporate world is hardly a rationale for intimating that upper-level white-collar crime is unremarkable—especially for a perspective that claims to be interested in the etiology of crime.

Equivocations concerning numbers—in which some are held to provide a direct route to reality, while others obscure it—are not accidental. The contradictory assumptions concerning figures run parallel to the drawing of racial distinctions and typologies. The differential evaluation of numbers ensures that Black Americans are rendered an object in need of biosocial criminological explanation and intervention. The criminal behavior of whites, in comparison, has no place in biosocial criminology; it has been transformed into an "unmarked" category. Because figures establishing the "disproportionality" of white criminality cannot be found, it fails to qualify as a problem and is thus excluded from the biosocial gaze. Moreover, any significance that the criminalized behaviors of elite whites may have for interrogating biosocial criminological theory is lost. This is inexplicable for a perspective that purports to decipher the mysteries of crime by recourse to genes, brain patterns, and evolutionary imperatives irrespective of political context.

Adaptive and Maladaptive Adaptations

This brings us to our final equivocal moment within biosocial criminology, which concerns the notion of "adaptation." To unpack the problem here, it is necessary to provide a brief overview of the biosocial criminological model for theorizing "crime" and/or "criminality" and its indebtedness to Charles Darwin's arguments concerning evolution.

Biosocial criminology seeks to distinguish itself from biological criminology by abandoning the notion that "biology is destiny." As critics and proponents have noted, the revised mantra of biosocial criminology is "nature via nurture."[100] In other words, it is posited that nature—a signifier for the body, its genetic material, its mind and brain, and how it has come to be via evolutionary history—provides a grid of possibilities, some of which will be evoked or brought to fruition by environmental forces. In this view, genes, brain structure, evolutionary histories, and so forth are not posited as programmatic causes of particular behaviors such as "crime." Instead, biology provides the individual with "traits" or general "faculties," such as "aggression," "impulsivity," or "risk-seeking." The concrete form and expression that these "traits" come to take is contingent upon how biological material interfaces with the environment. The environment is construed as promoting specific behaviors from among the biological grid of possibilities on the presumption that such behaviors facilitate survival and reproduction.[101]

This account of human behavior owes a fairly obvious debt to Darwin and his *Origin of Species*, which posits that the behaviors and strategies of biological organisms are "adaptive."[102] In the realm of nature, organisms are essentially experimenting with how best to calibrate themselves to the physical environment that sustains their existence. Organisms are biologically primed to try out adaptations: those that facilitate reproduction are likely to be maintained for as long as they remain functional; those that do not will become dormant or disappear. To be sure, organisms that do not adapt and reproduce will eventually die off. Importantly, Darwin generally spoke of this natural order

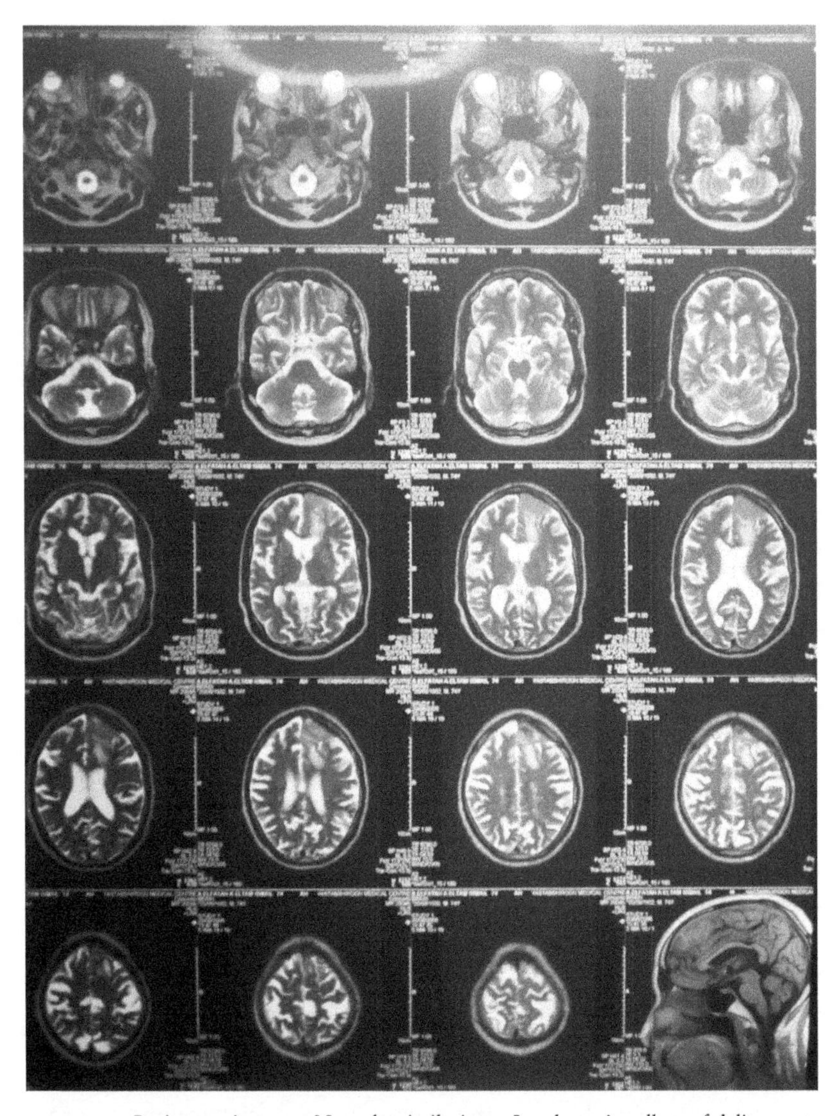

FIGURE 4. Brain scan imagery. Note the similarity to Lombroso's gallery of delinquents (cf. figure 2). While many may now be skeptical that the human face reveals criminality, the view that activity levels in the brain reveal psychopathy is gaining traction. *Source:* Alamy, "MRI Scans of a Patient's Brain" (CRNFRY), n.d.

as transpiring without rhyme or reason: one struggles in vain to find within it a grand plan, design, or ultimate purpose.[103]

"Adaptation" is a tricky concept within Darwin and evolutionary logic. Within such a framework, any manifest behavior is presumptively adaptive, and this forecloses the possibility of such behavior being maladaptive. It would be hard for biosocial criminology to claim otherwise given that it conceptualizes behaviors as being pulled from a biological substratum by environmental conditions. If one were to assume oneself equipped to decipher whether a behavior is adaptive or maladaptive, it would be necessary to explain why the interactions between nature and the environment have generated a maladaptive behavior. In other words, a "maladaptive adaptation" is oxymoronic according to the axioms of evolutionary logic.

To be sure, it might be possible to construe some behaviors as maladaptive in retrospect. Maladaptation might be signified by a species dying out to the point of disappearance.[104] Prior to such an event, however, there is no yardstick by which one can presciently identify maladaptive behavioral expressions. Any such estimation would amount to an interpretive exercise. And of course even the suggestion that death signifies maladaptation is interpretive given that natural history is devoid of any underlying purpose or design. There is simply no way to know how things were "meant to be," so the death of a species might just as easily be "adaptive" in some broader conceptualization of the natural order.

Another way to construe maladaptation as meaningful—and this would appear to be the preferred strategy of biosocial criminology—is to posit that environments may change while adaptations developed in the past continue to manifest in behavior.[105] In this scenario, the biological substratum somehow finds itself out of sync with the environment. This position, however, contradicts the very principle—"nature via nurture"—that is said to distinguish biosocial criminology from "older" versions of biological criminology. For traits grounded in biology to manifest in behaviors that are disharmonious with the environment presupposes *biological determinism*, not *biological-environmental interaction*. How else can one explain that the environment has suddenly ceased to facilitate translating general biological dispositions into particular or manifest behaviors?

Despite these problems with the concept, biosocial criminology routinely claims that some behaviors—most obviously "crime," but sexual practices are also frequently discussed—can safely be classified as "maladaptive." And once such behaviors are designated as maladaptive, the biosocial criminologist is free to judge them according to moral and political sentiments. Alternatively, one might say that to classify a behavior as maladaptive cannot but be a moral, political judgment—a move that is logically barred by evolutionary logic and by biosocial criminology's own efforts to claim a zone of "pure science."

The equivocal moment transpires when behaviors that are theorized as *to be expected* in light of evolutionary logics and/or the "nature via nurture" principle are simultaneously condemned on normative grounds. "Adaptation," in other words, frames all behavior as adaptive when it is a matter of sanctioning the supposed objectivity of biosocial theory, but adaptations become maladaptive when the biosocial theorist wishes to evaluate behavior from a normative, political standpoint. "Adaptation" thus renders "crime" normal *and* abnormal, natural *and* unnatural, adaptive *and* maladaptive. The same equivocation engulfs biosocial criminology's discussion of sexual behaviors, especially evident in the obsession with r/K mating strategies. Here, r strategies are said to involve men sleeping with as many women as possible in order to ensure the continuity of genes. By way of contrast, K strategies involve men sleeping with few women and remaining within the domestic setting to raise offspring. The r strategy is associated with Black communities and is used to explain single-parent households and crime; the K strategy basically codifies the normative "nuclear family" model in evolutionary language.[106] The former is normatively condemned, the latter taken as politically desirable, even though both constitute "adaptations."

By way of further illustration we can return to Walsh's account, unparalleled in its excellence for revealing the contradictions of biosocial criminology: "Far from being pathological in a naturalistic sense, the behavior [criminalized activity, violence, sexual promiscuity] of inner-city males fits exactly what evolutionary biologists would expect given the cultural situation in which they find themselves."[107] Nevertheless, the very next page provides assurance that "this does not mean that such behavior is necessarily adaptive in modern environments."[108] And in a moment that effectively conjoins the two senses in which adaptation (adaptive and maladaptive) is used: "African American males living in honor subcultures may be responding to their situation in ways designed by nature. This is not an exculpatory statement; immoral behavior is not excused by its natural (i.e., designed by natural selection) origins."[109]

How biosocial criminology can maintain such a position without dissolving into a mess of logical contradictions is beyond us. The "inner-city male" behaves in ways that are adaptive—the behaviors are described as an "exact fit" for the "cultural situation." And yet we are then instructed to regard this perfect fit as maladaptive, somehow out of place with the environment. As with previous equivocations, the incompatible meanings attributed to adaptation make sense when biosocial criminology is read as a political discourse. To paint racialized others as criminal by nature, the biosocial criminologist is dependent upon "adaptive adaptation," but the notion of maladaptation is necessary to justify political intervention.

This could be taken as further support for Carrier and Walby's reading of biosocial criminology, in which it figures as a Ptolemizing of Lombroso. Equivocating on adaptation ultimately constructs Black Americans as Lombroso's "atavists": their behavioral adaptations may be a perfect fit for some bygone era, but now render them imperfect for the modern world, if not "out of place." This raises an obvious dilemma: What can be done with "matter" that is purported to be "out of place"?[110] Far too often in human affairs, this "problem" is resolved via incapacitation, quarantine, eugenics, or genocide. Biosocial criminology demands refusal of this point, which is grounds for the refusal of biosocial criminology.[111]

CHAPTER 3

Actuarial Science

Crime Control as a Risky Business

The fundamental premise of actuarial science is that future events can be predicted on the basis of aggregate data. The aggregated data can include measures of any number of variables, and almost any kind of event can be the object of prediction. Correlations among variables are sought, then used to estimate the probability of future events. One of the most familiar examples of actuarial assessment in practice is the provision of car insurance. Among the most important events to insurance companies is the accidental car crash. Gathering data on accidents allows insurance companies to develop a sense of how frequently they occur and what driver characteristics are associated with accidents. The costs of insurance will vary according to individual characteristics and their power to predict the likelihood of being involved in an accident. In this context, actuarial risk assessment is problematic, but the stakes are relatively low. For example, insurance costs for young men are slightly higher than for other reference groups because they present a greater risk of being involved in an accident.

Much like car insurance, the use of actuarial logic within criminal justice contexts is problematic. However, criminal justice does differ in that it is an institutional arena in which the stakes are very high. This in itself should make us cautious about engaging in forecasting and prediction within this domain. Nevertheless, the logic of actuarial science has increasingly pervaded criminal justice systems over the last one hundred years or so.

Arguably, actuarial science made its first forays into criminal justice in the United States around the 1930s via probation.[1] Since then, it has been used to develop "criminal profiles" intended to preemptively detect offenders, assist in bail decisions, rationalize the imposition of different types of sentences for individual offenders (e.g., rehabilitative versus punitive sentences), and determine lengths of incarceration. The logic of actuarial assessment can also be discerned in particular strategies of punishment, such as preventive detention and "three strikes and you're out" legislation. As this list intimates, sometimes the use of actuarial logic is formalized and rigorous, but it can also be utilized in more rudimentary ways. When used in probation and sentencing decisions, actuarial assessments are much more likely to involve the use of a tested risk assessment instrument.[2] In other contexts, however, predicting future events is done in a rather crude manner. "Three strikes" legislation, for example, essentially treats an individual's criminal record as an infallible predictor of future criminality.

The allure of risk prediction is not difficult to understand. If it is possible to predict when crimes will occur, or which individuals will commit criminal acts, then offending can be apprehended before it happens. In its purest form, this is the "precrime" of *Minority Report*.[3] Surely, so the argument goes, there can be no rational objections to such an endeavor. However, many problems accompany actuarial logics and techniques, and these are especially pronounced when utilized within criminal justice contexts where individual freedoms and liberty are at stake.

The critique that follows is directed toward actuarialism irrespective of the particular contexts in which it is rationalized and utilized. In other words, problems with risk assessment surface whether the technology is employed in policing, sentencing, or making parole decisions. With reference to particular examples and illustrations along the way, our critique revolves around three interrelated themes.

First, we argue that actuarial science is not fully understood if juxtaposed with principles that are often held to underpin criminal justice, such as rehabilitation or retribution. In such views, actuarialism and its prioritizing of risk are often construed as embodying a profound departure from rehabilitative ideals or penal welfare. Alternatively, some suggest that actuarialism and rehabilitation—especially insofar as the latter seeks to correct the moral attitudes and behaviors of individuals—have merged, thereby producing a hybrid mode of criminal justice. We do not necessarily disagree with much that is found within this literature.

However, what we suggest is that actuarial science—rather than existing within an orbit that includes rehabilitation or retribution—harbors within its own boundaries a logic of sacrifice. Sacrifice suggests that *homo sacer*—the life that can be killed with impunity, as Giorgio Agamben puts it—is the point of application for the mathematical or statistical calculations of actuarialism. In this sense, actuarial science may appear reducible to a series of mathematical calculations, but these are animated by a profoundly "nonscientific," almost romantic, notion.

Following from this, we argue that actuarial science conceals the logic of sacrifice by which it is animated via narratives that reduce the enterprise to questions of "accuracy." In such narratives, it is often presumed that if risk assessment and prediction instruments can operate with some degree of accuracy, they can be regarded as meeting or surpassing ethical concerns. This, however, glosses over the ways in which sacrifice entails victimization and practices that would readily be framed as "criminal" were they to transpire outside the remit of state authority and violence. That is, if the practices instantiated by actuarial science within criminal justice were orchestrated by private, nonstate actors, it is difficult to see how they would not invite criminalization.

In the third part of this critique, the nonrandom distribution of sacrificial victims is emphasized. To put this otherwise, sacrifice is socially patterned. It corresponds to, and thus reproduces, fundamental power asymmetries and inequalities. We suggest that the intersection of actuarialism and politics is discernible in several ways.

First, rather than simply identifying objective risks, actuarial technologies are dependent upon the encoding of specific norms or templates of normality. As Sasha Costanza-Chock shows, departures from these norms are flagged as anomalous and then construed as risks, which means that not all individuals are subject to predictive techniques and algorithms in the same way. Some bodies are more adversely affected than others.[4]

Second, and closely related, actuarial prediction cannot be reduced to a value-free judgment premised upon a clear demarcation of "independent" or "predictor variables" and "dependent variables." Drawing from critiques of "white logic" and how it infuses criminological research and criminal justice practices, actuarialism can be said to construct and impart specific meanings to the variables that it positions as independent.[5] For example, it is not so much the case that "unemployment" predicts "future crimes," but that positing such a relationship construes an unemployed individual as a threat.

Third, we argue that the politics of actuarialism are evident in its choice to treat whatever correlations might be found within aggregate data as pertaining solely to individuals, rather than social relations. In other words, actuarial data could just as easily be understood as mapping the various degrees of threat that structural problems pose for particular individuals or social groups, but this possibility is rarely—if ever—entertained by proponents of risk assessment. In making this argument, we do not wish to return to a realist account in which actuarial data and their correlations are taken as evidence that structural problems are the cause of criminal behavior. What we are suggesting, however, is that actuarial data could be narrativized in at least two ways, but only one of these narratives is generally regarded as self-evident. The narrative that has emerged as "second nature" asserts that risk assessment has implications only for individuals. This narrative—in which risk assessment represses any significance that it may have for social structure—tells us much about the politics embedded within actuarialism: it is a technology that seeks to inscribe criminality on society's most marginalized members such that gross inequalities are preserved, if not exacerbated.

POLARITIES OF THE ORTHODOX: REHABILITATION VERSUS INCAPACITATION, DISCIPLINE VERSUS CONTROL

Much of the debate concerning the rise of actuarialism revolves around its relationship to rehabilitation, or the "penal welfare" model that dominated punitive policy for much of the twentieth century. Alternatively, retribution may be utilized as a counterpoint to prediction. Once these punitive logics are drawn into comparisons and understood as central to theorizing the punitive field, it is not uncommon to find that at least one of them will be valorized, almost as if such valorization were a precondition for critique of the other.[6] In some cases, the focus shifts to how the polarities have fused. Both modes of analysis, however, appear to block recognition of how a particular concept, and the practices that it instantiates, is central to actuarialism: sacrifice.

According to the rehabilitative ideal, or "penal welfare," punishment ought to focus on "reforming" the individual, typically by (re)calibrating them to dominant moral values and understandings of "normal" subjectivity. The notion of retribution conceives of punishment as "payment" for criminalized behavior. In its idealized version, retribution construes punishment and crime as two components of a transaction.[7]

While the cost of punishment may slightly exceed the benefits from crime, their exchange is presumed to restore balance or bring closure to the harms of criminalized behavior. Rehabilitation and retribution are similar in that they are retroactive; they respond to individuals according to the criminalized behavior that they have purportedly engaged in.

In contrast to these "responsive" logics, actuarialism is governed by preemptive control of future risks. Not surprisingly, it is therefore typically associated with incapacitation, which attempts to render the commission of criminalized behaviors physically impossible. There is little concern with remoralizing individuals or exacting payment in this model of punishment. Instead, the emphasis is on suppressing behaviors that it is presumed will transpire in the future if incapacitation or prior interventions are not utilized.

While these polarities—rehabilitation or retribution versus preemptive strategies and incapacitation—often structure assessments of actuarialism, there is disagreement concerning exactly how they are related.

From Morally Oriented Rehabilitation to Cold Prediction

According to Malcolm Feeley and Jonathan Simon and David Garland, actuarialism and the rise of risk logics embody a significant departure from rehabilitative ideals.[8] Feeley and Simon situate the rise of actuarialism within neoliberalism, a form of political economy known for intensifying social precarity and producing an ever-growing "surplus population," positing that it has generated a need for distinct punitive logics and discourses. They suggest that actuarialism—or the "new penology"—is discernible in new discourses, objectives, and techniques.[9]

The discourses of the rehabilitative ideal posit that individuals ought to be the locus of state-orchestrated interventions that will "correct" problematic dispositions and behaviors. Individuals are still present in the "new penology" but are no longer understood in the ways presumed by rehabilitation. As Feeley and Simon put it, individuals remain, "but increasingly they are grasped not as coherent subjects . . . but as members of particular subpopulations and the intersection of various categorical indicators."[10] The correction of souls has been replaced by the management of people in light of risk categories and probabilities of future offending.[11] In addition to recrafting how the penal subject ought to be imagined, actuarial justice inflects constitutional norms and protections with new meaning. For example, "probable cause" has been turned on its head, such that it is now possible to target individuals in light of generic profiles grounded in aggregate data, rather than requiring specific reasons to suspect an individual.[12]

In terms of objectives, rehabilitation is geared toward crime reduction by reintegrating individuals into society. In comparison, the new penology is concerned with identifying and managing groups that are extraneous to neoliberalism. Crime is no longer thought of as a social problem that can be eradicated. Instead, the new penology assumes that crime problems will inevitably persist and, as such, the best we can hope for is their management and containment.[13]

Concerning techniques, rehabilitation provides individuals with skill sets that facilitate community integration and law-abiding behavior upon release from prison. Conversely, when the goal is to contain and manage crime, there is little need for techniques premised upon reintegration. Prisons become "no frills" warehouses that simply incapacitate those who are thought to be dangerous. Moreover, individuals who are released but end up returning to prison come to be read as evidence of effective penal policies; if people are returned to prison, then the institution is "fulfilling" its obligations to contain those deemed "dangerous" or "high risk."[14]

Like Feeley and Simon, Garland accepts that criminal justice systems are best understood as intimately connected to general social conditions, and that Anglophone countries underwent profound structural transformations in the post–Second World War era. Garland expresses the transition as one from "penal welfare" to "cultures of control." For much of the twentieth century, as Garland tells the story, economic conditions generally improved in ways that benefited populations in the Anglophone world. By the 1950s, labor was in demand, and this ensured job stability, high employment rates, and wages that allowed one to support a family while maintaining a good standard of living. The role of the state included regulating the economy to protect basic living standards. Such economic conditions ensured a society that was relatively cohesive, with low crime rates. Most of society's members could be provided with incentives that encouraged acceptance of social rules and expectations.[15] Given the relative comfort of the period, a sense of concern and duty could be extended toward others, even those who were less fortunate and turned to crime. Psychological and sociological understandings of crime and punishment were in vogue. Both disciplines encouraged the view that engaging in criminalized behaviors was somewhat beyond the control of individuals, but that the underlying causes of crime could be addressed through rehabilitative interventions.

By the late 1960s and early 1970s, the social order that underpinned penal welfare was in crisis. The core driver of penal change was, arguably, economic globalization or the intensification of communication. Developments in communication meant that companies could relocate their manufacturing capacity to the so-called third world for the sake of exploiting cheaper labor power. Many manufacturing jobs in the United States and the United Kingdom vanished within a very short time frame. These economic shifts altered social relations. The 1970s and 1980s saw increased levels of family breakdown, higher crime rates, and greater animosity between the poor and those who were relatively well off.[16] The power of the state was surpassed by large economic players, and this usurping of authority had implications for social life; the state's ability to regulate social problems, especially that of crime, was greatly diminished. The growing power of capitalism corresponds to a weak, "limited sovereign state."[17]

Discourses well attuned to the new social order swelled and gathered ground. There arose the "nothing works" approach to punishment. In this view, rehabilitation efforts are seen as ineffective and thus futile. Somewhat related, the idea that some people are "simply evil" experienced a resurgence during the 1970s and 1980s. Furthermore, a range of discourses that emphasize how crime can be inhibited emerged. For example, "situational crime prevention" suggests that criminal offending is a product of opportunities that are available within everyday routines and the nature of physical space. From here it follows that preventing crime entails organizing social and physical space such that it cannot be enacted.[18]

The penal field that corresponds to these sociocultural developments, according to Garland, amounts to a "culture of control." In this form of penality, the state increasingly outsources crime control to a series of nonstate or quasi-state actors: private security firms, business improvement districts (BIDs), community watch groups, and individuals. At the same time, state punishment focuses on incapacitation strategies that exclude individuals from society for prolonged periods of time, if not permanently.[19] This component of the penal field is facilitated by actuarial science or risk assessment.

Garland—and, albeit to a lesser extent, Feeley and Simon—takes much inspiration from Foucault's view that discourses emerge against a backdrop of broad social developments and instantiate practices that adjust individuals to social demands. In focusing on reformist discourses, which promised to "correct the souls" of offenders, Foucault revealed how they were damaging to freedom. Such discourses promised that some greater good would follow from rehabilitative interventions—for the individual offender and the social order—but they ended up crafting penal systems that saturate everyday life and permeate subjectivity such that individuals conform without being aware of the subtle, manipulative mechanisms by which this is achieved.[20]

When actuarialism is said to follow upon some supposed crisis in rehabilitation, however, the perniciousness that Foucault saw in the latter appears to fall by the wayside.[21] And this seems to promote a dichotomous logic in which rehabilitation is equated with "good" (or moral, ethical), while control is equated with "evil" (or amoral, unethical). That "penal welfare" is something of an oxymoron appears to be forgotten. This is probably most obvious in Garland's description of the industrial era, in which people were employed, nuclear families prevailed, crime rates were low, common morality dictated that we should be concerned with one another's welfare, and the state provided security. One may certainly speak of social changes over the period, but such an account of the penal welfare era tends to construe it as some kind of "golden age" to which we ought to return. It is difficult to square this with Foucault's critique of reformist discourses. Then again, perhaps it says something about actuarialism if it can make disciplinary regimes appear a better alternative!

The Freedom of Risk: Souls on the Reservation

Pat O'Malley rejects the idea that risk and, by implication, actuarialism, can be understood as a totalizing logic. This entails rejecting any suggestion that penal fields and criminal justice systems have been overrun by some monolithic conceptualization of risk.[22] For O'Malley, risk cannot amount to a totalizing logic because it is devoid of meaning in itself. There is, in other words, nothing "essential" about risk.[23] Instead, how it comes to be understood and refracted in practice is dependent upon the political ideologies in which it becomes enmeshed. If, for example, risk combines with the political logic of neoliberalism, less punitive approaches may be favored. However, if neoconservative political ideologies take control of risk-based approaches, then punitive policies are likely to follow.

O'Malley suggests that once risk is understood as devoid of essence—or as politically neutral—the concept's critical potential can be discerned. This critical potential—in O'Malley's view—is especially evident in neoliberal appropriations of risk because supposedly these are not concerned with the morality of individuals in the way that neoconservative logics are. If Feeley and Simon and Garland secretly endorse a return to rehabilitation, O'Malley comes to valorize risk and actuarialism for their supposed respect for individuals and their subjective desires.

Take, for example, O'Malley's endorsement of "control." Such strategies are seen as giving individuals the freedom to roam as they see fit, provided they do not stray beyond the parameters of Gilles Deleuze's "reservation": "Risk has an enormous potential for tolerance. Whereas discipline is focused on closed institutions and shaping individuals into conformity with a norm, risk technologies may simply work to police security boundaries—leaving the space within relatively open and morally floating."[24] Rather than totalizing, risk now figures as "tolerant" because it abandons moral discipline. And instead of acknowledging the "pessimism" in Deleuze's notion of "societies of control," not to mention Foucault's "governmentality," O'Malley welcomes it, transforming it into a source of "optimism." In our view, however, this can only work by forgetting that within logics of control one is hemmed in by classification schemas, the codes generated by data doubles, risk instruments, and so on. In other words, O'Malley more or less concedes that those narratives in which risk is understood as totalizing are not without foundation; moral license there may be, provided that its exercise remains within a space effectively totalized by risk. By this logic, one may as well argue that the "no-frills" prison is a site of freedom and tolerance because it is more or less unconcerned with how incarcerated individuals think and what they might do to one another.

Furthermore, and as Loïc Wacquant's account of neoliberal statecrafting indicates, O'Malley misconstrues the relationship between neoconservatism and neoliberalism. Rather than divergent political logics with some points of overlap, they are better understood as two sides of the same coin. As Wacquant argues, what O'Malley regards as two political logics that ultimately stand in tension with one another appear as parts of a bigger whole when understood in light of contemporary social relations.[25] Neoliberalism, which extends and intensifies the social relations and inequalities that mark capitalism, pushes many into precarity and insecurity in order that a small portion of the population can accumulate excessive wealth. To maintain this state of disequilibrium, the draconianism of neoconservatism is summoned. Neoliberalism may respect the individual, but this depends upon your place in the social order; neoconservatism may be authoritarian, but this also depends on your place in the social order. As Wacquant puts it: "Penalization is not an all-encompassing master logic that blindly traverses the social order. . . . On the contrary: it is a skewed technique proceeding along sharp gradients of class, ethnicity, and place, and it operates to divide populations and to differentiate categories according to established conceptions of moral worth. At the dawn of the twenty-first century, America's urban (sub)proletariat lives in a 'punitive society,' but its middle and upper classes certainly do not."[26]

It seems unlikely, then, that risk is as malleable or plastic as O'Malley argues.[27] Whatever malleability it may have is unlikely to exceed the social relations that demarcate the neoliberal state. As such, to argue that appropriating the "neoliberal" understanding of risk moves us away from "pessimistic," totalizing views, especially those espoused by the likes of Feeley and Simon, amounts to recuperating the neoliberal order, which now appears a bastion of freedom for its purported departure from an authoritarianism that meddles with the individual.

Amalgamations of Risk and Discipline: Nominal Skepticism and Hybridization

To be sure, it is possible to dispute the notion that risk can be understood as a totalizing logic without apologizing for neoliberalism. Robert Werth's work on risk assessment within parole illustrates this possibility. Werth finds that upper-level managers encourage parole officers to use risk assessment instruments. While frontline workers follow the directives of management, they remain skeptical that risk instruments alone can provide a complete understanding of an individual. Much of this skepticism can be attributed to protecting the professional standing of parole officers: if a risk assessment instrument that almost any individual can fill out and interpret is sufficient, there is little justification for treating parole decisions as a specific form of expert judgment.[28] In seeking to preserve the professional status of their occupation, parole officers remain dependent upon "clinical," individualized modes of assessment, thereby generating a "hybrid" mode of assessing individuals and reaching decisions concerning parole.

Despite Werth's "resistance thesis," which problematizes readings in which risk figures as a totalizing logic, this does not lead to the view that risk can be construed as a site of critical potential. This is perhaps most evident in Werth's argument that risk technologies entail specific "performative effects."[29] In the parole context, risk instruments effectively construct all individuals as "risky." Granted, one may be classified as a "low," "medium," or "high" risk, but it is not possible to be classified as "zero risk." This last category is never a possibility, according to risk assessment instruments.

All individuals assessed by a risk instrument, then, will be constructed as representing some degree of danger. This is the "performative" side of risk—the moment that the risk instrument makes a statement that does not necessarily mirror an external reality but actively produces the risky subject. And according to Werth, this performative moment ultimately provides ideological support for mass incarceration, supervision, regulation, and the criminalization of large portions of the population. This effect seems to follow despite the political ideology that engulfs risk.[30]

Kelly Hannah-Moffat also identifies how the logics of risk and individual assessment merge. Much of Hannah-Moffat's work focuses on how power asymmetries inform hybridizations of risk and reform. For example, examining how parole boards use risk assessments in determining whether imprisoned women will be released back into the community, Hannah-Moffat shows that although they may be familiar with the contexts in which women's offending often occurs, this is elided by risk instruments because they isolate discrete behaviors and treat such moments as the basis for making judgments about an individual's rehabilitation and future behavior.[31]

Many women are imprisoned for violent acts directed against a male partner, but this typically occurs only after experiences of men's violence and, in numerous cases, histories of past abuse. That is, most women who commit a violent act against male partners do so in self-defense. But only the violent act is recorded by the risk assessment instrument and, moreover, given license to position the incarcerated woman somewhere within a schema of risk categories. In order for women to improve their likelihood of parole, they are expected to show that they have successfully "rehabilitated" themselves via the completion of anger management programs, demonstrating "insight" concerning the "triggers" of their violence, accepting ownership for their actions, and so on. Engaging in these kinds of narrative performances is likely to be effective insofar as they may alter perceptions of how much risk one poses if released. There is, however, a rather profound sense in which responsibilizing women for their past actions and their future behavior is unjust. The hybridization of risk and rehabilitation tasks women with self-monitoring their thoughts and proposed courses of action; the expectation is that women will absorb the problem of men's violence by coming to act as their own moral guardians.[32] Insofar as this is the case, it is not men's violence that is constituted as a risk to women, but women's acts of self-defense that are regarded as a threat to society.

In a slightly different context—that of algorithmic facial recognition technologies—Claudio Celis Bueno pursues a line of thought that is similar to Werth's and Hannah-Moffat's, emphasizing how the power of aggregate data returns to the individual, thereby merging with disciplinary power. Algorithmic face recognition is being developed and implemented in various settings. The technology's presence can be discerned when social media platforms ask if an individual user wants to "tag" a photo, but it is also surfacing in areas such as "border patrol, targeted advertising, and police profiling."[33]

Algorithmic face recognition entails the production of massive databases that contain thousands, possibly millions, of facial images. As an element of a database, the facial image is "pre-personal" and "supra-personal."[34] In both of these respects, the technology is unconcerned with the individual. It is "pre-personal" in that the face is broken up into "bits of information that no longer belong to a private individual." At the same time, it is "supra-personal" insofar as these "bits of information are collected as part of a large machine . . . that searches for predictive patterns."[35] However, it is such databases against which an image of any particular face can be searched for a match, or against which a face can be categorized in some specific way. In these respects, the technology can construct an individual subject as a "potential 'consumer,' 'criminal,' or 'terrorist.'"[36]

It is in this capacity to link the "pre-" and "supra-" personal to the individual that multiple modalities of power intersect. As Bueno argues, in algorithmic face recognition, "the face plays out two semiotic regimes: an asignifying machine which connects the deterritorialized elements of the intensive face with the deterritorialized flows of information fed to the machine-learning algorithm; and a signifying machine that reterritorializes these flows on the reflective face and the private individual."[37] Put more simply, algorithmic face recognition operates at the intersection of control (those moments of deterritorialization, or asignification) and discipline (those moments of reterritorialization, or signification). Rather than being contradictory, or representing two modes of power that belong to distinct temporal moments, these "diagrams of power" are better understood as working together in a mutually reinforcing manner.[38]

From Totalizing Logics and Hybridization to Sacrifice

In the work of Werth, Hannah-Moffat, and Bueno, neither risk, rehabilitation, nor retribution is necessarily valorized, and the emphasis on "hybridization" is important for reasserting that the individual remains the site of intervention when risk techniques are utilized.[39] However, the emphasis on hybrid modes of governance, or intersecting diagrams of power, appears to miss precisely how the individual subject resurfaces in risk logic. In our view, the problem is not so much that moral training (discipline) and actuarial prediction (control) merge, but that a logic of sacrifice is endemic to the latter. In this sense, the actuarial never jettisoned individual subjects only to return to them armed with disciplinary techniques. Rather, the actuarial is premised upon identifying those who can be sacrificed.

To put this otherwise, the actuarial needs to be understood not so much as a mathematical or statistical departure from the ideals of rehabilitation and/or retribution, but more as fundamentally wedded to a logic of sacrifice. The actuarial does not summon its power by portraying itself as an advance over the allegedly failed logics of "discipline" or "penal welfare," nor does its power stem from its promise to control behavior. Furthermore, it is not fully understood if it is thought to follow from situating itself at the intersection of two modalities of power, a "deterritorializing" algorithm that ultimately "reterritorializes" the individual subject. Rather, buried within the core of the actuarial, one finds the harboring of a very old, prescientific, religious sentiment.

Søren Kierkegaard's *Fear and Trembling*, an exploration of the story of Abraham and Isaac, provides one possible avenue into the notion of sacrifice.[40] Abraham is told by God to take Isaac, his only son, to Mount Moriah and offer him as a sacrifice. Abraham does so willingly, keeping God's command a secret even from his wife, Sarah. As Abraham is about to kill Isaac, thereby proving his faith, the angel of God intervenes and prevents fulfillment of the sacrifice. For Kierkegaard, the parable elucidates how faith is of a higher order than the ethical. This is so because Abraham is incomprehensible: killing one's son is the gravest of sins, yet he is prepared to do it because God has ordered it. Were Abraham to try to explain his actions to another, however, he would be regarded as unfathomable, bizarre. By way of contrast, ethical decisions may involve difficult choices, but such choices can be rationalized and meaningfully explained to others.[41]

FIGURE 5. *Abraham's Sacrifice*, etching by Rembrandt, 1655. Various scholars focus on the relationship between incapacitation and retroactive punishments, such as rehabilitation or retribution. Incapacitation, however, cannot but entail sacrifice. *Source:* Image courtesy of Princeton Museum Art Museum, Laura P. Hall Memorial Collection.

In Kierkegaard, faith is regarded as of a higher order than the ethical precisely because it cannot be rationalized. The "knight of faith" acts according to a "law" that emerges independently of the sociocultural realm. Religion is a source of such autonomy, and is thus valorized by Kierkegaard. Sacrifice is the litmus test of one's faith. We suspect that had God commanded the actuarial to take Isaac to the mountain, the child would most certainly have been slaughtered. In any case, the actuarial aligns itself with Abraham: incomprehensible from an ethical perspective but justified by a faith that is imparted from beyond the social. It is not surprising to find, then, that advocates laud this leap of faith; critics posit its unfathomability.

The notion of sacrifice is also central to Agamben's account of modern political order (democracy and totalitarianism). Unlike Kierkegaard, Agamben does not read the will to sacrifice as an indicator of faith or autonomy, but as the principle upon which modern political power rests.[42]

For Agamben, sacrifice is wedded to the notion of *homo sacer*, which refers to the individual who can be killed with impunity. Sacrifice can be literal or metaphoric. In either case, those sacrificed are understood as included-excluded, or as included via their exclusion from the social. Agamben suggests that the concentration camp summarizes the predicament of *homo sacer*: those inside the concentration camp are excluded from the social, but they are simultaneously within the clutches of the state. And in this space of "inclusion-exclusion," anything can be done to them: they can be compelled to participate as subjects in brutal scientific experiments, tortured, and mass murdered. Hence, they are sacrificed, killed without being killed, killed with impunity.[43]

Homo sacer and its inexorable potential to be sacrificed are not peculiar to the concentration camp. As noted, *homo sacer* is symbolic of a particular model of power, one that Agamben elucidates by drawing a contrast with Foucault's "panopticon" or disciplinary power. Rather than being invested in the production of docile subjects, *homo sacer* suggests that power resides in the capacity to decide who is to be relegated to spaces of inclusion-exclusion—where anything can be done to the individual—and to enforce such decisions with impunity. It is the state that assumes such power and authorizes itself to wield it as it sees fit. Akin to discipline, all are subject to the possibility of sacrifice, and so any individual or group could be rendered *homo sacer*.[44] The theory is terrifying.

One does not need to accept wholesale Agamben's attempt to theorize the formation of modern states to see the points of contact across *homo sacer* and the individuals who find themselves in the crosshairs of actuarial science. Actuarial science takes *homo sacer* for granted and reproduces this diagram of power in its very logic and its numerous applications.

In other words, the actuarial does not simply move from "discipline" or "retribution" to "control," nor can it be reduced to an amalgamation of "old" and "new" punitive strategies; it is a means to determine which individuals can be sacrificed. In the moment that one becomes the subject of an actuarial decision, they are "innocent," but their "guilt" is projected somewhere in the future or into a zone of probability—a place where it cannot materialize or be represented (or "demonstrated"). And yet individuals can be imprisoned, electronically monitored, detained, subjected to cognitive reprogramming, and so on.

Advocates of the actuarial seem to acknowledge that it is sacrifice—not discipline—that surfaces as the conceptual twin of forecasting. Somewhat ironically, it is not necessarily on mathematical or statistical grounds that the endeavor is rationalized; rather, justification is said to reside in the will to produce the included-excluded, a moral-political choice. Jean Floud, for example, notes: "It is likely that at least two persons are detained for every one person who is prevented from doing serious harm. . . . [There is] at least a 50 per cent risk of being unnecessarily detained."[45] Given the significant rate of error in prediction, the "knight of faith" must be brought to the fore and must be prepared to ignore the "angel of God" that prevents sacrificing the innocent. Floud continues:

> The justification [for preventive detention] can only be that we are thereby relieving someone else of a substantial risk of grave harm—that we are justly redistributing a burden of risk that we cannot immediately reduce. . . . We have to make a moral choice between competing claims: the claim of a known individual offender, not to be unnecessarily deprived of his liberty; and the claim of an innocent (unconvicted), unknown person (or persons), not to be deprived of the right to go about their business without risk of grave harm at the hands of an aggressor. . . . Providing the risk is real and the anticipated harm is grave, the fact that it is diffused is no argument, prima facie, for refusing to take preventive measures.[46]

Despite admitting that prediction techniques cannot actually tell us when the "risk is real and the anticipated harm is grave," some will simply have to be sacrificed for what is purported to be the greater good: the right of "innocent persons" to go about their business in the absence of "aggressors." It never seems to occur to proponents of the actuarial that the predictions they so fetishize also constitute a known threat to "innocence." Detention is sanctioned despite the knowledge that it is—in all likelihood—unjustified. To get out of this problem, previous convictions are conflated not with "guilt" in its common legal sense, but with a conception in which it signifies a "loss of all innocence" that is permanent and unassailable. For the Nazis, being Jewish was grounds for sending one to the concentration camps; for advocates of the actuarial, previous convictions essentialize the individual and provide the warrant to escort one to zones of inclusion-exclusion.

"ACCURACY," "COST RATIOS," "SECURITY": ERASING THE SACRIFICIAL VICTIMS OF ACTUARIALISM

The actuarial strives to conceal the sacrificial victims that it cannot but produce by reducing itself to an available technology and, in so doing, limiting interrogation to questions concerning its "accuracy" and instrumental efficiency. In repressing its will to sacrifice, it silences what could be described as its own "criminality" or production of victims. The discursive eluding of victimization is most evident in the pervasive drawing of comparisons to "clinical judgment," efforts to neutralize the problem of "false positives" (manifested, for example, in the necessity of choosing a "cut-off score" or ratio of false positives to false negatives), and guarantees to engineer public security.

Error and Ethics Both Start with E: "Surpassing" Clinical Assessment, Being "Accurate Enough," or the Elixir of "Cost Ratios"

In "clinical" judgment, an assessment of an individual's dangerousness and likely future behavior is made. This is done by an expert—generally a psychiatrist—and is typically based on a range of psychological criteria that are nonactuarial in nature. The expert psychiatrist typically regards "dangerousness" as a "yes" or "no" problem: an individual is deemed to be either dangerous or not dangerous. It is often claimed that this leads clinical judgment to overpredict dangerousness.[47] Dependency upon a subjective or interpretive process can lead experts to err on the side of caution, especially in criminal justice contexts. No psychiatrist wants to risk their professional credibility or being sued for stating that an individual who proceeds to offend was not dangerous.[48] Moreover, clinical judgment is held to be relatively crude given that it assesses individuals in categorical rather than continuous terms.

Actuarial judgment also entails assessing the threat that an individual presents and their likely future behavior, but the assessment is made on the basis of aggregate data that enable the positing of correlations among predictor variables and behaviors. The individual is scored on particular criteria and, on the basis of this score, assigned somewhere within a schema of risk classification. In this, actuarial methods treat the individual as an example of the group to which they are assigned by the risk assessment instrument. Where one is located in relation to a variety of subpopulations will have implications for which courses of action ought to be adopted.[49]

It is often thought that actuarial instruments keep the subjective tendency to be overly cautious in check because they are premised upon correlations found among variables within large data sets. In this, the actuarial appears as a form of "mechanical objectivity": the product of a machine that operates without subjective predispositions. Moreover, it speaks in terms of probabilities, likelihoods, and chances. In positioning individuals somewhere along a continuum of risk, actuarial scientists are never really "wrong" in their assessments. After all, they produce a likelihood of future behavior, not a "yes/no" proclamation.[50]

In the literature on prediction, it is not uncommon to find a narrative in which the probabilistic and mechanical aspects of actuarial judgment render it accurate relative to clinical (and other modes of) judgment.[51] Once the relative accuracy of actuarial methods is posited, their use is often regarded as acceptable.

However, and as intimated earlier, the problem with this view is that actuarial methods are still liable to significant error, particularly when used in criminal justice settings; the types of criminalized behavior that generally concern the actuarial are rare, making their accurate prediction a complicated, difficult task.[52] Nevertheless, the advance that actuarial assessment supposedly embodies relative to clinical judgment is chalked up as a rationalization for its incorporation into criminal justice settings. Somewhat ironically, it is better because—unlike clinical judgment—it is *not* excessively cautious.[53] The possibility that neither method should be given much weight—especially in light of high error rates or inaccuracy—seems to receive scant attention once comparisons are drawn between the two. That the actuarial seems to "perform better" is construed as evidence of its moral legitimacy and inevitability.[54]

The problem of "false positives" is inherent to prediction and is amplified by the need for actuarial models to select a "cut-off score." Any particular actuarial prediction will fall into one of four possible categories: a "false positive," "false negative," "positive hit," or "negative hit." A false positive occurs when an individual is predicted to fail but would in fact succeed. In a sentencing scenario, a person would be incarcerated even though they would not have offended if left in the community. False negatives involve predictions of success accompanied by actual failures. In other words, a false negative amounts to releasing someone who later offends. A positive hit transpires when one is predicted to fail and does in fact fail (continuing to detain an individual who would have offended if released, thereby preventing some crimes). Finally, negative hits occur when one is predicted to succeed and does in fact succeed (releasing someone from prison who does not proceed to offend).[55]

Implementing a predictive model necessitates choosing a "cut-off score," or the number of false positives to be tolerated for a certain amount of false negatives. However, and as Stephen Gottfredson and Don Gottfredson show, the difficulty is that reducing false negatives (i.e., allowing individuals to remain free who then offend) inevitably entails increasing the volume of false positives (i.e., detaining individuals who do not pose a threat).[56] The ratio of false negatives to false positives is not a trivial problem. We have already made reference to Floud's admission that in order to detain one individual who would have offended if left free, it is likely that two innocent people will need to be detained.[57] But it is worth emphasizing that when actuarial judgments are used in parole and sentencing contexts, the decisions often implicate *years* of additional incarceration.

Gottfredson and Gottfredson are quite thorough in exploring the technical aspects of prediction and, on this basis, intimating how the incorporation of actuarial methods into criminal justice settings is fraught with problems. Nevertheless, they ultimately summarize the problem that actuarialism presents as one of better scientific accuracy such that false positives will be minimized: "Unless predictive accuracy can be increased, reducing false negatives can only be done at the expense of increasing false positives."[58] In addition, "prediction in criminal justice settings clearly is not sufficiently accurate to form the basis of social policy."[59] The assumption being made, it would seem, is that if false positives can be reduced, or made to stand in inverse proportion to false negatives, then the use of actuarial methods would be less problematic as a basis for social policy. In other words, implementing actuarial techniques is sound based on the proviso that they are "sufficiently accurate."[60]

There is an irony to this logic. The indispensable condition for improving the accuracy of actuarial predictions is more data points. Such data points, however, are effectively counts of the very problems that actuarialism promises to address. To predict interpersonal violence with greater accuracy, for example, it is necessary to have more cases of such violence and more data concerning its correlates. In other words, the fetish for predicting future behavior presupposes that some level of the behaviors that actuarialism readily accepts as "criminal" be maintained. In this sense, the technical requirements for making prediction more accurate cannot be squared with the claim that actuarial techniques offer a viable mechanism for controlling "crime."

Whereas Gottfredson and Gottfredson suggest that prediction needs to reach a specific degree of accuracy before its use in criminal justice can be entertained (i.e., the volume of false positives needs to be strongly curtailed), others posit that the false positive problem can be dissolved by simply dropping it in the economic waters of "cost ratios."

Richard Berk, for example, notes that criminal justice officials generally regard false negatives as more concerning than false positives, and that officials "treat false negatives as far more costly than false positives. Costs ratios of ten-to-one, or even twenty-to-one, are common."[61] Curiously, these ratios appear to be a direct inversion of William Blackstone's famous legal formulation: "It is better that ten guilty persons escape than that one innocent suffer."[62] The phrase "or *even* twenty-to-one" intimates that criminal justice officials are inclined to adopt extreme ratios, yet Berk notes: "[There] may be concerns that the cost ratios imposed have little empirical justification. . . . On the one hand, that may not matter. What matters are the *preferences* of stakeholders, and a key feature of these preferences can be captured in elicited cost ratios."[63]

In plain English, the inaccuracy problem can be dismissed by asserting that each instance of a false positive costs much less than a false negative. According to this logic, "stakeholders" could justify any gross imbalance between false positives and false negatives by developing a "preference" for ratios in which the cost of a false negative always exceeds (or matches) whatever number of false positives it corresponds to. If the ratio of false positives to false negatives is 100 to 1, the number of erroneous predictions can be washed away by asserting that the cost of a false negative would be 100, whereas the cost of a false positive is 1. There may be "little empirical justification" for this 100-to-1 cost ratio, but that does not matter because it is, after all, a matter of arbitrary preferences determined by stakeholders.[64]

As Berk would have it, the volume of false positives can be understood as a choice that criminal justice officials can be entrusted to make. Error is eradicated by assigning an arbitrarily determined cost to false imprisonment. In this cost determination, expressed in the currency of releasing a "risky" individual, false imprisonment is more or less guaranteed to be "undervalued." We saw a similar logic surface in biological crime science, where some erroneous findings of guilt could be regarded as inconsequential and thus dismissed. In actuarialism, we find the same cavalier attitude when the freedom of the "other" is at stake.[65] This cavalier approach is only exacerbated given that false positives and false negatives tend to produce very different reactions within criminal justice settings. "In a situation where there are dire consequences to missing a true positive (i.e., not identifying a high-risk offender as such) and few direct costs to officers when making false positives (i.e., wrongly identifying a low-risk offender as high-risk), it is easy to see why officers would yield to the so-called precautionary principle."[66] By no means is Berk alone. Disregard for the freedom of others surfaces quite frequently, albeit in various guises, among those with a predilection for prediction.[67]

Promises of Security: Major Reductions in Crime, Minor Prison Populations

The desire for actuarial instruments with a relative degree of "accuracy"—or for concepts that can absolve such instruments of their insurmountable inaccuracies, thereby legitimizing their use—is intensified by the potential benefits often attributed to them. In early formulations of using prediction in sentencing, for example, it was argued that crime rates and the number of those imprisoned could be reduced if criminal justice systems were able to identify "habitual offenders," or those individuals who supposedly commit a disproportionate number of crimes. If this were possible, prisons could be reserved for that small portion of individuals who repeatedly offend, and their selective incapacitation would ensure lower crime rates.[68] This amounts to claiming that actuarial techniques will control "major risks," thereby ensuring the public's protection and security.

According to some accounts, however, the use of risk assessment and prediction—especially in the contexts of policing and punishment—may actually lead to increases in general crime rates.

In Bernard Harcourt's view, one of the major limits to actuarial science stems from "elasticities" in offending. Elasticity can be understood as the extent to which patterns in offending will be altered when risk prediction is incorporated into criminal justice practices. The logic of prediction tends to assume two primary actors: those who offend and those who police offenders. However, Harcourt suggests that we should be concerned with three actors: those subject to profiling, nonprofiled groups, and those tasked with policing.[69]

Recognition of multiple players begins to pry open the limits to prediction. To create a criminal profile (or risk classification schema, etc.) presupposes that specific factors are associated with a particular type of offending. However, if there is a strong correlation between some group characteristic and a type of offending behavior, it is highly likely that reasons exist for such a relationship. In the case of transporting drugs, this may well be driven by exclusion from labor markets and thus economic precarity. But if this is so, then it is unlikely that creating a criminal profile of the drug courier will lead to decreases in transporting drugs. To fix that, we would need to reduce asymmetries in economic opportunities. Without addressing these macrolevel dynamics, the underlying factors that drive the criminal behavior will outweigh concern with the possibility of being profiled and possibly arrested. For marginalized individuals, profit still outweighs risk: the targeted group will have a "low elasticity" in terms of their offending behavior, and as such, profiling will not alter behavioral patterns.

As noted, though, there is more to Harcourt's story. The use of predictive techniques implies the presence of nonprofiled groups. What is likely to happen to their behavioral patterns given that they do not match criminal profiles or "high-risk" groups? Somewhat ironically, prediction seems to hinge on rational choice theory in making sense of criminalized behavior (i.e., greater surveillance of profiled groups should increase the cost of offending and thus inhibit it), but pays little attention to nonprofiled groups. It is difficult, however, to avoid the conclusion that offending rates are likely to increase among those from nonprofiled populations. The focus on lone Black male drivers creates a ripe opportunity for carpooling white grannies carrying the mother load of Bolivian marching powder. After all, criminal profiling dictates that such groups, precisely because they do not embody the attributes associated with offending, do not merit police scrutiny. In this scenario, why not offend if your chances of being caught are significantly reduced? There are, of course, serious implications to all of this: it is unlikely that profiling will reduce offending insofar as profiled groups are resilient. Furthermore, if profiling provides those from nonprofiled groups with greater incentives to offend, then it is not unlikely that overall crime rates will increase.[70]

Whereas Harcourt emphasizes that individuals from nonprofiled groups will be incentivized to commit crime, others have suggested that this can also hold for those from profiled groups. "Three strikes and you're out" legislation exemplifies this argument. "Three strikes" is a crude appropriation of prediction logics, premised on treating past offending as an infallible basis for gauging future behavior. In places where it has been introduced, there is a tendency to rationalize three strikes in light of its supposed deterrent value. As the argument goes, if only criminals knew they would receive life sentences or be made to serve the full term for a third conviction, then they would think long and hard about reoffending.

Despite the aura of plausibility that surrounds such an argument, studies have thrown it into question. Thomas Marvell and Carlisle Moody, for example, find that "three strikes" is associated with increases in homicide.[71] This finding has been replicated in a subsequent study by Tomislav Kovandzic, John Sloan, and Lynne Vieraitis.[72] As counter-intuitive as such findings might sound, they are not difficult to understand. Although there are regional variations, three-strikes legislation typically specifies a range of "serious crimes" as strikable offenses. In the version of the law that was enacted in California, for example, strikable offenses included violent crimes, such as homicide, but also incorporated property crimes, such as burglary.[73]

Therein lies the incentive to homicide. An individual who has two strikes recorded against their name presumably knows that they risk a life sentence if caught for homicide, but they also know that the same penalty applies for every other listed offense, such as burglary. The legislation effectively treats qualitatively distinct offense types as equivalent for the third-time recidivist. Given this, why not take offending behavior to the extreme of homicide if the penalty will be the same as for burglary? This is an odd way to put the problem perhaps, but it is not facetious. In accounting for increases in homicide after the introduction of three-strikes laws, Marvell and Moody, as well as Kovandzic, Sloan, and Vieraitis, posit that an offender may be more inclined to kill witnesses of a strikable crime, or police officers who attempt to make an arrest, to avoid a conviction that will lead to a life sentence.[74]

The Insurmountability of Sacrificial Victims: "Crime Rates by Omission" and the Paradox of "Perfect Accuracy"

The critiques offered by Harcourt, Marvell and Moody, and Kovandzic and colleagues are premised upon a fairly conventional understanding of "crime rates," not to mention a realist notion of crime. A handful of scholars, though, point out that selective incapacitation presupposes a particular understanding of "crime rates" and how they ought to be tallied. As John Blackmore and Jane Welsh note, one of the assumptions upon which selective incapacitation rests is that crimes committed in prison can be ignored.[75] Todd Clear and Donald Barry identify the same problem but provide a slightly different formulation when they state that incapacitative policies operate with a "target-specific model": "If an incarcerated thief continues to steal from fellow prisoners, we normally consider the thief to be incapacitated nonetheless. However, this target specification becomes more troublesome when the 'incapacitated' offense is, for example, homicide."[76]

For Clear and Barry, homicide provides the strongest warrant to question whether "selective incapacitation" can actually be said to incapacitate: "When incarceration results in prison homicide, for example, it is a bit harsh to suggest that 'incapacitation' has occurred when in reality all that has occurred is that the population of potential victims has been limited to convicted felons and their guards."[77] Taken together, critiques focused on actuarial promises to reduce crime posit that prediction may actually lead to higher crime rates, but even if this does not transpire, crime may remain relatively stable despite appearing to decline. This is because incapacitative strategies relocate crime to areas where it occurs, yet can elude being tallied. The first claim, articulated by Harcourt (and extended in a particular context by Marvell and Moody and Kovandzic and colleagues), does not acknowledge the notion of a "crime rate by omission," whereas it plays an important role in the second argument, advanced by Blackmore and Welsh and Clear and Barry. To be sure, Harcourt's argument could easily integrate the notion, as it would only embellish the fundamental point: predictive techniques will not reduce crime.

In any case, Clear and Barry convincingly posit that advocates of selective incapacitation require a particular conceptualization of the "crime rate," one in which various omissions can be regarded as nonproblematic. And it is fair to say that glossing over these omissions is necessary insofar as doing so shores up the claim that crime rates can be reduced by utilizing risk assessment instruments during sentencing.

It also follows from Clear and Barry's argument that—strictly speaking, and notwithstanding the mysterious weight accorded to homicide—"incapacitation" and "crime rates by omission" have as their logical corollary the impossibility of criminality within prison, and thus the impossibility of victimization in prison. This in turn carries with it the additional propositions that an absolute distinction between "offenders" and "victims" can be drawn, and that once an individual is essentialized as an "offender" (especially if this triggers incarceration), it no longer makes sense to understand any of their experiences within criminal justice as constituting victimization.[78]

In our view, however, the notion of a "crime rate by omission" needs to be taken a little further. Not only does the actuarial assert its effectiveness by refusing to acknowledge behaviors that it would most certainly add to the crime rate—and treat as important measures within its risk assessment instruments—were they to transpire beyond the prison's walls, it also fails to acknowledge that any course of action based on its predictions cannot but mean the production of victims, all of whom will most certainly elude being counted or tallied in any way whatsoever. In short, practical actions born of actuarial prediction—were they not orchestrated by the state—would in all likelihood readily be construed as "crimes with victims."

That actuarialism is inconsistent concerning the conditions according to which it will count identical behaviors as criminal and as indicators of risk is not surprising. The inconsistency is logically indefensible, but it is necessary to justify selective incapacitation, preventive detention, and so on. Behaviors commonly rendered as illegal and transpiring *outside* of prison are always to be coded as "criminal," as this constructs the objects of actuarialism, thus justifying its associated practices. But those same behaviors are to be ignored if they occur *inside* the prison or, more specifically, once the "risky individual" has been incapacitated. Ignorance and willful omission are necessary to ontologize "incapacitation," to make it seem as though there is no gap between predictive techniques and their purported effects to "neutralize" risky subjects. And least surprising of all, the practices instantiated by actuarial assessment are never to be evaluated against available criminal codes. That would reveal their own congruence with behaviors commonly deemed and treated as illegal.

That a preemptive decision of the false positive variety—such as erroneously incapacitating an individual via imprisonment, denying a "reformed subject" parole, detainment on the basis of a profile, and so on—amounts to a form of victimization is not a difficult argument to make. If such a preemptive decision leads to incarceration, the individual is being kept in prison for an act that they have not committed and would not have committed if released. In the case of detaining on the basis of a criminal profile, the individual's constitutional rights have been violated insofar as there exists no grounds for individual suspicion.[79] If a private actor were to preemptively incarcerate another person, it is not unlikely that this would be framed as "kidnapping" or "false imprisonment"; if they were to "detain" on the basis of some profile and search someone's pockets, frisk them, and so on, this might easily amount to "assault" or, at the very least, fairly strong evidence of being an asshole.[80]

But even if incapacitation followed from a prediction instrument that somehow achieved perfect accuracy—that is, guaranteed correct identifications and thus restricted to only those "false negatives" and "positive hits" who would have offended if left at liberty—this would still entail victimization. In such a scenario, the individual who finds themselves in such circumstances has not committed a criminalized act, has not been found guilty of a criminalized act, and has not been sentenced in light of a guilty finding. Indeed, given that the incapacitation is based on a prediction of future behavior, the criminalized act for which one is detained cannot have happened and therefore cannot have been adjudicated. As with "false positives," if the state was not the dominant actor in this scenario, the preemptive action would most likely attract the label "false imprisonment."

Various scholars acknowledge that state intervention in the absence of a criminalized act (or reasonable grounds for suspicion, and so on) poses a problem. It is not uncommon, however, to see preemptive decision-making framed as a question of whether it can be reconciled with other principles of punishment, such as just deserts, fundamental legal principles, or individual rights.[81] To our knowledge, very few scholars have pointed out that were something like preemptive detention to transpire beyond criminal justice—or beyond the state's "monopoly of violence"—there is a very strong likelihood that it would be read as "criminal." The tendency to refrain from positing the state's criminality can perhaps be understood to follow from regarding its exercise of power as fundamentally legitimate, rather than as a declaration of the right to determine the exception. This would explain why some construe preemptive strategies as problematic insofar as they may threaten the perceived legitimacy of the state.[82]

Among advocates of actuarial justice, arguably the most brazen strategy for wiggling out from under the weight of this dilemma—that is, sanctioning interventions and/or punishment before history unfolds—is to admit that preemptive action is unjust, but to insist that it will "protect the public": the "eternal return" of "security." According to Douglas Cunningham, "The public deserves to be protected. . . . If selective incapacitation is a kind of post-conviction preventive detention, so be it."[83] As Blackmore and Welsh indicate, utilitarian arguments along such lines offer a formula that purports to resolve the dilemma: "the public must be protected even at the cost of some injustice."[84]

Such a formulation, however, hardly provides a sound rationalization for preventive practices, given its performative effects. The formula positions security and justice as irreconcilable; that both can be obtained is rendered impossible. From here, injustice is reconfigured as a fair price to pay for protection. But if this idea is pushed to its logical limit, no one is protected, and security cannot be realized. The formula exonerates the denial of any and all human rights, provided this is done in the name of security. Various moments from the US-led war on terror provide ample illustration of the practices that follow when security is construed such that it necessitates and exonerates injustice. Guantanamo Bay, for example, is not incongruent with this defense of preemptive action, but rather its very embodiment. And as Richard Ericson notes, it is "painfully evident that the vast majority of those detained as possible terrorists post-9/11 are false positives."[85] Gross injustices are neutralized in the name of security. Preventive detention, selective incapacitation, profiling, and so on will be found somewhere along the same continuum as such state-orchestrated atrocities.

The formula in question is better read as a confession of sorts, an acknowledgment that injustice constitutes the indispensable condition for actuarial science. And yet this injustice is presented as the mechanism that exonerates the actuarial and the forms of insecurity buried within it. The political standpoint that animates pitting security against justice—in claiming that injustice is necessary for security—becomes obvious when juxtaposed with alternative formulations that might also be entertained, such as "justice may require some degree of insecurity" or "the pursuit of justice and security ought to be one and the same."

SELECTIVE INTERPRETATION: POLITICAL MOBILIZATIONS OF ACTUARIAL DATA

Thus far, we have suggested that actuarialism may appear to be a mathematical enterprise, but that it harbors an inner logic of sacrifice. We have also posited that acknowledgment of its sacrificial victims is repressed by deflecting debate to some of the technical issues posed by prediction. In this final section, attention is drawn to the identification of *homo sacer* and how this process is anything but random. That is to say, actuarialism transpires in political contexts, which makes it much more likely that individuals and groups from particular regions of social space will find themselves ensnared by its technologies and applications. As with crime science writ large, actuarialism inscribes criminality on particular bodies while simultaneously refusing to inscribe it on social structures and institutions.

Power asymmetries and prediction can be understood to intersect in at least three ways. First, because they embed assumptions about what constitutes "normal" and equate this with "nonthreatening," actuarial technologies will target and single out those individuals who depart from their encoded norms. Departing from norms, however, is not necessarily the same as constituting risk. Second, and closely related, prediction does not simply reflect some level of risk posed by particular individuals or groups, but actively constructs some as risky relative to others. Actuarialism then amplifies that sense of risk by targeting specific groups. Third, it could be said that the decision to reduce actuarial data and their correlations to a mechanism for distributing individuals according to risk classifications is in itself political. The aggregate data upon which actuarialism relies—the seedbed for its "discovery" of correlations among variables—could just as easily be read as evidence of social, rather than individual, problems. If such data were understood to map structural problems, something like "social relations" would become the object that poses a "low," "moderate," or "high" level of risk for particular individuals or groups. In other words, the use of aggregate data to construct some individuals as posing danger to the social is entirely arbitrary, a political choice in no way dictated by some objective reality. There is no empirical or logical reason not to invert the order of things that actuarialism takes for granted, thereby treating aggregate data as a mechanism for sorting social arrangements according to the level of risk they pose for specific individuals and groups.

Templates of "Normality," Constructing the Risky Subject, Mining Risk

In analyzing the politics of airport body scanners, Costanza-Chock provides a clear illustration of how predictive, algorithmic technologies are not objective or neutral in their application, but, echoing Lombroso, loaded with assumptions concerning what constitutes a "normal" subject. Consequently, they operate in discriminatory and biased ways. Not all individuals are equally grist to the mill of risk assessment. Describing the experience of approaching and walking through an airport body scanner, Costanza-Chock writes: "If the agent selects 'male,' my breasts are large enough, statistically speaking, in comparison to the normative 'male' body-shape construct in the database, to trigger an anomalous warning and a highlight around my chest area. If they select 'female,' my groin area deviates enough from the statistical 'female' norm to trigger the risk alert, and bright yellow pixels highlight my groin. . . . In other words, I can't win."[86]

Cis-normativity is built into the airport body scanner. As a result, "Queer, Trans, Intersex, and Gender Non-Conforming (QTI/GNC) folks" are invariably read as anomalies and thus "disproportionately burdened by the design of millimeter wave scanning."[87] Unlike those who fall within the "normal" parameters of the body scan templates, they will be constructed by the technology as "matter out of place" and thus subject to greater levels of surveillance and scrutiny.[88] Along with Costanza-Chock, Shoshana Magnet and Tara Rodgers have shown that it is not only QTI-GNC folks who are likely to find themselves subject to greater policing. The same holds for people of color, those with specific religious affiliations, people with disabilities, and those situated at the intersection of attributes deemed to deviate from norms.[89]

The example might seem idiosyncratic, but the core point is broadly applicable. Rather than being premised upon some kind of "pre-" or "supra-social" design, one that might be imagined to wash over individuals like waves wash over sand at the beach, risk technologies produce risky subjects by presupposing templates of "normality" or conformity.[90] From here, the second way in which actuarial techniques and power asymmetries intersect emerges. The experience recounted by Costanza-Chock demonstrates that the body does not amount to an objective risk that is simply detected by the screening device. Rather, the screening device inscribes risk on some bodies. To push this a little further, concurrent with the capacity to identify the anomalous, the scanning and sorting aspects of risk attribute particular meanings to departures from its encoded range of "normality."

Several scholars have drawn attention to the problem of "white logic," a notion that encapsulates how white privilege is embedded within criminal justice practices and criminological research, thereby reinforcing and supporting racial stratification. Kathryn Henne and Rita Shah, for example, argue that white logic is especially discernible when one considers how "race" is incorporated into research projects. As they note, race tends to be construed in one of three ways by researchers: "When studies in our sample do analyze race, they often narrowly configure it as a variable that essentializes racial categories. This emerges in three ways: (1) "White" was the most oft-used reference category; (2) race was coded as an independent variable; and/or (3) race was coded as a moderator or control variable."[91]

In the first conception, "white" is assumed to designate the group that embodies normal behavior, such that other racial groups are rendered deviations from the norm. The second conception typically divides race into "white" and "nonwhite" (or creates a limited number of categorical possibilities) and assumes that there is something about falling into these categories that explains behavior. Similarly, when positioned as a moderator variable, it is implied that "race changes the level of the effect under review."[92] It could be said of all three conceptions that they construe race as static and thus reproduce the logic of biological determinism.[93] Moreover, accounts that operate with such limited conceptualizations render it more or less impossible to consider how race and racism work in much more profound ways to structure the very phenomena under study.

Framing the problem as one of "colorblindness," Nicole Gonzalez Van Cleve and Lauren Mayes bring this line of critique to bear more closely upon criminal justice and its increased reliance upon actuarialism. In their view, "apparatuses and policies of criminal justice" need to be understood as a "set of social practices that constitute race and thereby impact racial identities, racial perceptions, and the myriad of complex meanings attached to racial categories."[94]

In the context of the "new penology" and its fetish for risk instruments, this implies that the very process of assessing the degrees of risk that individuals pose constructs the meaning of what are purported to be "predictor variables." In other words, it is not that things like "unemployment," "youth," "previous criminal history," "drug use," and so on amount to neutral, robust predictors of "criminal behavior," but that criminality comes to be associated with these factors by the risk instrument. Insofar as the predictor variables in risk assessment instruments are "race-embedded variables," the meaning of race is also constructed via actuarial techniques.[95]

Positing "criminality" (or "danger," "threat," etc.) as the signified of "unemployed," "unmarried," "youth," and other predictor variables is not without consequences. Harcourt summarizes the problem well in suggesting that actuarial science entails a "ratchet effect." In using this notion, Harcourt seeks to capture the ways in which profiling and predictions of future behavior intensify the socially precarious position of those most subject to such logics.[96] This follows from dependence upon predictor variables that turn out to be proxy measures—Van Cleve and Mayes's "race-embedded variables"—for constitutionally protected social characteristics, such as "race," class, or gender.[97] To return to "unemployment," which often surfaces as an important component of a risk assessment instrument, this is very much connected to one's class location; "criminal history" never fails to appear, but this very much correlates with racial marginalization.

Using such measures to identify "at risk" populations legitimizes concentrating resources on specific segments of the population. In directing resources in such a manner, however, one is more or less guaranteed to "discover" a greater portion of "risky subjects" in the population under surveillance, intensify the punishments of those accorded higher risk scores, and so on. As Harcourt shows, the ratchet effects of prediction shed further light on why those from racially marginalized social groups are overrepresented in the prison population of the United States and, among other things, why we find concentrated disadvantage in specific neighborhoods.[98]

Developments in predictive policing also demonstrate how actuarial technologies construct particular individuals and groups as "risky subjects," thus exposing them to a greater degree of state intervention. Predictive policing is something of a spectrum, with crude versions at one end and more technologically enhanced forms at the other. Racial profiling on motorways and "stop-and-frisk" tactics—especially well known in New York City—provide examples of relatively crude techniques. The growth of "big data surveillance," however, signifies a relatively recent trend within predictive policing. In this modality, the goal is to assign specific individuals to risk categories and use such classifications as a basis for surveillance.

In examining how surveillance based on big data has altered practices within the Los Angeles Police Department (LAPD), Sarah Brayne notes that predictive policing (of the big data variety) works by gathering a wide array of information and storing it in searchable databases.[99] The LAPD, for example, begins by recording when and where crimes occur. Following this, individual-level data from areas with high crime rates, or high rates of a particular type of offending, are gathered and logged. This may include information gleaned from "patrols, the Parole Compliance Unit, field interview (FI) cards (police contact cards), traffic citations, release from custody forms, crime and arrest reports, and criminal histories to generate a list of 'chronic offenders.'"[100] Once high-crime areas and "high-risk" individuals are known, such knowledge can be used to inform the allocation of police resources. It almost goes without saying, but those assigned a high-risk score become the object of greater police surveillance.

This may appear to be an objective process, one that approximates an impartial, "scientific" approach to policing. However, any database that identifies high-risk individuals or helps to generate a "criminal profile," then proceeds to direct policing interventions, is socially produced over time and thus embeds social relations. In the case of the LAPD and its use of big data, Brayne notes that a disproportionate amount of policing resources are concentrated on predominantly Black and poorer communities.[101] Indeed, discriminatory policing practices have been noted by many scholars in other contexts.[102] Of central importance, historic and contemporaneous police practices, despite their lopsided and skewed nature, are inevitably fed into the "predictive policing algorithm."[103] This is a fairly obvious point, and we have had reason to note it in the context of biological crime science: databases cannot magically generate the information that they end up storing; they are not immaculate. Instead, the contents are available because of humans who, operating within particular sociocultural frameworks, gather and input information.

Predictive policing registers citizen contacts with police officers and the criminal justice system but does not bother to record the politically and socially discriminatory contexts in which such contacts occur. The individual who ends up constructed as a risk, and thus subject to intensified surveillance and greater chances of police intervention, has little—if any—ability to question or resist this process. The only option left to the individual is to absorb the impacts of discriminatory policing, which now appear unproblematic, if not justified, because it seems as though a computer algorithm told police officers whom to surveil. As Brian Netter argues, risk assessment and prediction ultimately saddles "individuals with the burdens of generalizations on the groups to which they belong."[104]

Fetishizing Risk: Treating Individuals as a Threat to the Social, Rather Than the Social as a Threat to the Individual

This brings us to a final point, which concerns the arbitrary meaning that is attributed to actuarial data. That actuarial data are brought to bear upon individuals or particular groups—rather than social relations—reveals their inseparability from power relations. In developing this argument, we turn to Marx's analysis of commodity fetishism, which has come to occupy an iconic position within social theory. According to Marx, the commodity was fetishized in that it seemed to magically appear, as if commodities could safely be regarded as objects without a past. Commodities, however, must have a history, a series of social processes that make it possible for them to appear in the marketplace. To only see the commodity, but not the complicated social process that precedes it, is to fetishize, to confuse a small part for the whole. It was the commodity's "backstory" that Marx thus investigated.[105]

Commodity production is premised upon specific social relations. Any given commodity requires the capitalist to set in motion some form of coordinated labor power among individual workers. Furthermore, the capitalist will assess the productivity of this labor power in light of the socially averaged time it takes to make a commodity. Thus, if making a pair of boots generally requires five hours of coordinated labor power, a capitalist will perceive a problem with workers if they exceed this duration in producing boots. In this sense, socially average labor time operates as a structural force that plays a major role in governing the production process.

As this description intimates, the capitalist possesses all the power in this relationship. It is they who set the production process in motion and assess its overall efficiency. It is a staple of Marxist thought to posit that the power of the capitalist stems from ownership of the means of production. But this power works to benefit the capitalist in another important way. Insofar as they orchestrate the productive process, capitalists retain control of the commodities produced by the labor process. This suggests that while the laborers produce everything, they ultimately hand over the fruits of their labor to the capitalist. The capitalist takes the commodities to market and, via their exchange, converts them into money. Not surprisingly, however, the profit secured through this process is not redistributed among the workers, but rather is retained privately by the capitalist. This allows the capitalist to preserve, if not further entrench, their dominant social position. As such, commodity production hinges on the exploitation of labor, but it also secures privatized profits that are wielded against workers, thereby ensuring their continued subjugation.

Marx does not oppose the use of labor power and the development of our collective abilities to make things that could enhance our existence. Indeed, commodity production and the technological developments that underpin it are capable of benefiting society. The problem is that this is not how capitalism utilizes our productive capacities. As Marx saw it, capitalist modes of production militate against a fair distribution of the wealth that they create; they are not interested in raising the basic living standards of every member of society, as this would necessitate a relatively small capitalist cohort having to sacrifice some of the luxuries it enjoys.[106]

But what has any of this got to do with actuarial science? In our view, the logics that Marx identified in capitalist modes of production parallel those within actuarial science. The fetishism that engulfs the commodity is akin to the fetish one finds attached to the main object of actuarial science: the risk assessment instrument. And much like the commodity, a series of social relations and processes could be said to precede the production of the risk assessment instrument. Actuarial science, however, renders invisible these social relations by directing its gaze toward the individual and the risk they supposedly embody.

Certain structural arrangements and social problems can be read into the production of risk assessment instruments. For example, a significant portion of the population needs to be unemployed and to engage in "crime" to discover a correlation between these variables; similarly, many individuals need to be convicted, sent to prison, and released only to reoffend in order to discover that previous incarceration is a good predictor of future crimes. A certain level of family and community "disorder" can be presumed insofar as the actuarial scientist posits that this correlates with criminalized behavior. We could continue, but further examples would simply demonstrate the same point: any "predictor" that finds its way into a risk assessment instrument can be understood to sit somewhere between the social and the individual. If our political standpoint prioritizes the former, risk technologies direct our gaze to *social issues*.

To be sure, it could be said that reading actuarial science as a barometer of "social issues" logically follows from the fact that it is premised upon aggregate data. Social issues, in this sense, are very distinct from personal troubles. As C. Wright Mills put it: "When, in a city of 100,000, only one man is unemployed, that is his personal trouble, and for its relief we properly look to the character of the man, his skills, and his immediate opportunities. But when in a nation of 50 million employees, 15 million men are unemployed, that is an issue, and we may not hope to find its solution within the range of opportunities open to any one individual. The very structure of opportunities has collapsed."[107]

The distinction that Mills draws in this passage is of obvious importance in this context. Actuarial science inevitably maps patterns within a social order—yet lives by the narrative that its data pertain only to an individual's troubles. In this, the two are inverted, or perhaps one should say confounded, thereby transforming a social issue into a troubled individual. Marx's "wood table" certainly seems to be standing "on its head" and evolving from its "wooden brain grotesque ideas, far more wonderful than 'table-turning' ever was."[108]

Eliding the distinction between social issues and individual troubles guarantees that those who make possible the fabrication of correlations and thus prediction instruments—that is, those who experience the most intense forms of social marginalization—do not benefit from the knowledge that their structural experience generates. Aaron Doyle also notes that actuarialism translates social into individual problems, but takes this to mean that our approach to those deemed "risky" should be reconsidered. According to Doyle: "Marginalized people are exposed to more risks, but are also themselves categorized as bad risks: this constitutes part of the process of their exclusion. Thus, they are people who might need our help, but who are also, paradoxically, seen as a threat. . . . [M]ore and more in risk society, the emphasis is placed on the threat posed by the marginalized, rather than on helping them."[109]

This certainly highlights the paradox of actuarial risk assessment with much clarity, but formulating the problem in this way might be a little too paternalistic for some. Moreover, it tends to take for granted that contemporary social, economic, and cultural arrangements are generally acceptable but for their tendency to produce marginalization. In our view, it is not so much that individuals "need our help" per se—or that they need to be integrated into an order that is otherwise desirable—but that everyday life needs to be informed by an alternative range of discourses and practices.

If, for example, it is accepted that unemployment or precarious labor correlates with crime, this justifies efforts to ensure that full, meaningful employment is a priority of political economy. If previous incarceration is correlated with future offending, such knowledge justifies reflexive engagement with the prison as an institution and why it fails to reintegrate individual offenders. Or why not rethink punishment in a way that avoids excessive stigmatization of the individual? Perhaps knowledge of such a correlation implies that we should critically examine current legislation and institutional processes that criminalize and punish individuals, especially in the area of victimless crimes, such as personal drug use and dependency.

That the correlations found among variables that underpin risk assessment instruments are not used to critically reflect upon social relations verifies our broader point about technology fetishism being intimately tied to power asymmetries. Knowledge about risk could be interpreted as a mechanism for assessing the extent and intensity of social problems, or it could be reduced to a barometer of individual pathology. That those who end up in control of the risk assessment instrument adopt the latter approach arguably tells us more about the limits of our social, cultural, epistemic, and political frameworks than about the individual whose existence is determined by guesses about their future behavior.

It is, arguably, not surprising to find that risk logics in which the individual is the presumptive point of application have found strong footing within societies marked by drastic economic disparities and social precarity. Prediction promises "crime control," but its ultimate appeal resides somewhere beneath such surface appearances: to contain those individuals cast aside by structural arrangements without disturbing the latter.

CHAPTER 4

Security Science

Cartographies of Crime, States of Exception, and the Twilight of Liberty

Security involves the protection of persons or places: everything from a lone individual to an entire nation. Of course, traditional criminal justice organizations—such as the police—provide security of this kind, but they are usually excluded from the term. A different form of discourse is employed to describe their actions. The term *security* is typically used to designate actors from the private sector and/or the nation-state. In regions throughout the world, governments contract out security work to commercial providers. Governments and private security firms often form working partnerships in the hopes of providing security.

Security science may be understood to revolve around the identification of mechanisms and technologies that, if effectively implemented, promise safety. It treats the motivation of threats (offenders) as extrinsic, unknowable, and irrelevant, focusing instead on appointing capable guardians and on hardening vulnerable targets.[1] A variety of strategies and technologies are available—walls (physical and virtual), foot patrols, video surveillance, GPS, security checkpoints, alarms, motion sensors, and so on, often combined with the exercise of force—but security science remains indebted to a singular, overarching logic: the assumption that security is produced through the effective enforcement of perimeters. It is an exercise in liminality. That is to say, security science accepts as its fundamental goal the creation of bounded zones that are insulated from crime.

We develop our formal critique of security science by focusing on technologies that arguably embody the core practices associated with the production of security in much of the contemporary world. More specifically, our analysis draws from material on counterterrorism (e.g., the US Patriot Act, enhanced airport screening, drones, new techniques of communications surveillance), closed-circuit television (CCTV), and private security forces. The chapter revolves around four interrelated critiques.

First, we unpack the ontological and epistemological standpoint that underpins a great deal of contemporary security practice. After all, the underlying ontology and epistemology determines what is regarded as a security problem worthy of attention (e.g., terrorism) and what is not (e.g., capitalism); it informs conceptions of criminality and crime control, thereby governing how perceived problems are to be addressed. The epistemology of security is also key to understanding the odd constellation of criminalized behaviors that most security science attempts to regulate.

Second, we examine the costs of security. Not in the sense of a detailed economic cost-benefit analysis, but rather a sociological discussion of the mechanisms through which the public absorbs the costs associated with security technologies. Much of the privatized profits in the lucrative private security sector are derived from public, taxpayer funds. Insofar as this is the case, security technologies involve the funneling of public wealth into private hands. The close connections between governments and private economic interests explain some of this phenomenon. But if we wish to comprehend the reckless spending of public money on security, it is also essential to recognize society's culturally ingrained fetish for technological solutions to social problems.

Third, our analysis of security science lays bare this fetish for technology. It reveals an evangelical belief in panopticism, surveillance, and the promise of big data. But because government and industry adopt security tools without understanding their limitations, and because they employ tainted data, security remains an elusive goal. Technology operates as an addiction. When confronted with failings in security, policy makers do not reevaluate their reliance on the tool; rather, they double down and chase the next, better technological fix. Perhaps this is because, ultimately, the efficacy of the technology matters less than its mere existence. Technology has been enlisted not to redefine social relations between haves and have-nots, but to enforce them.

Fourth, we assess the relationship between security and the civil liberties ostensibly safeguarded by the state. Some critical scholarship interprets the democratic state as a protector of liberties and views novel security technologies as threats to those liberties. We argue, instead, that the modern state is a Janus-faced entity. In limited circumstances, the state honors constitutional frameworks that can protect individuals from the state's monopoly on violence. Miranda rights, Fourth Amendment search-and-seizure jurisprudence, and Eighth Amendment protections against cruel and unusual punishment are all well-known examples of such criminal procedure guarantees. But when managing undesirable conduct, the state can shift effortlessly from a framework of criminal jurisprudence to one of national security. Actions forbidden under criminal law become accessible when framed in terms of national security. And although individual actors might understand their behavior as circumscribed by the criminal law, it is the state, and the state alone, that possesses the power to invoke national security interests.

ONTOLOGY AND EPISTEMOLOGY OF SECURITY: CRIME, TERRITORY, AND RETERRITORIALIZATION

In practice, the core business of security science normally consists of protecting private property; protecting the interests, policies, and objectives of the state; and, to a lesser extent, controlling a narrow grouping of street-level offenses. Such practices hinge on specific ontological standpoints concerning which behaviors constitute "threats" or "criminal activity," and epistemological positions concerning how such behaviors are to be controlled. Security science presumes a criminal subject that can be kept at a distance by establishing boundaries or perimeters. There is little concern with "motivations" and possibilities of "displacement." Corresponding to this notion of the subject, security science cannot but endorse a territorial logic, whereby its agents are tasked with safeguarding a particular, delimited space, but need not be concerned with nearby areas. Furthermore, what might be described as a "traditional," hierarchical cartography is broken up and reterritorialized in light of what security technologies can observe and measure. This spawns a "horizontalization" of crime types, wherein innocuous behaviors are put on the same plane as behaviors typically regarded as warranting concern.

Instead of critically engaging the concept of "crime," security science—in the tradition of Beccaria and Lombroso—accepts as an objectively given reality the hegemonic constructions dictated by contemporary power relations. But as described of bio-forensic technology in chapter 2, "crime" and "security threats" do not possess ontological status. Rather, security science constructs these categories, rendering noise into signal.

If security science was truly data driven, then criminal justice systems might attempt to leverage this knowledge to eliminate what could easily be narrativized as the most pernicious forms of crime, such as white-collar and state crimes.[2] In terms of financial losses and wrongful deaths and injuries, the harms of white-collar crimes dwarf those of all aggregate street crimes.[3] State crimes, too, incur immeasurable harms.[4] However, the philosophy and technologies of conventional security science are particularly ill-suited to this kind of work.[5] They are intended to secure, not subvert, property interests; they are intended to maintain, not ameliorate, existing power asymmetries. It is therefore unsurprising that those whom C. Wright Mills called "the power elite" and whom Paul Fussell rightly called "the top out of sight" are not subjected to surveillance.[6] Indeed, those elites with access to charter aircraft never have to endure the long queues and humiliating security checkpoints that have become routine for the economy classes.[7] When these individuals are confronted by perimeter walls, alarms, cameras, and private security forces, it is because these systems are working on their behalf. Power enlists security "to protect and to serve" its interests.

The videotaped recording of the Rodney King incident in 1991, which seems like an obvious example of the legal phrase res ipsa loquitur (the thing speaks for itself), also illustrates how noise is processed into signal. The recording shows five white Los Angeles Police Department (LAPD) officers repeatedly tasering, kicking, and beating with their batons a prone Black defendant.[8] But when, in 1992, four of the officers were prosecuted in state court for using excessive force, all were acquitted.[9] Then president George Herbert Walker Bush marveled, "Viewed from outside the trial, it was hard to understand how the verdict could possibly square with the video."[10] Visual records might appear to be neutral and objective, but it is power that ultimately decodes their meaning.[11]

Alongside the assumption that hegemonic constructions of crime can be taken as self-evident, security science posits that criminal inclinations can be suppressed with technological controls, but not necessarily dissipated. In other words, it is presumed that criminalized behavior is energized or motivated, but its expression can nevertheless be blocked within the spaces bounded by security measures. Crime figures as a contingent act, or dependent variable, which can be inhibited by controlling its spatial correlates. Unlike environmental crime science, explored in the following chapter, this narrative does not necessarily go out of its way to discredit the notions of "innovation" or "displacement," which posit that criminalized behavior will remanifest in alternative strategies or seek out new sites for expression when repressed. It does not matter whether the threat is suppressed through deterrence or merely displaced; as long as the threat is eliminated within the secured perimeter, security science has achieved its objective.

Security science is reticent to denounce innovation and displacement because it exists within the logical tension between, on the one hand, requiring ongoing risks and threats (to justify its own existence) and, on the other hand, needing to appear competent in addressing and overcoming these threats. What is necessary, then, is an ever-present threat, albeit one that can be kept at bay. This justifies new innovations and technologies and perpetuates the life cycle of security science.

Constructing criminality and its containment in this manner is contradictory. To promise safety, security science assumes a criminal subject who will simply accept the boundaries established by its technological capacities: terrorists will cease plotting attacks due to enhanced airport security; homeless populations will cease loitering around private-public property because of security personnel; and thieves will desist from stealing because of alarms, fences, and so on. To justify its continual presence and fetish for innovation, however, security science holds that it cannot eliminate threats. This is arguably most obvious in the area of combating "terrorism," which involves massive economic investments in security technologies, frequently updating those technologies, and very rarely, if ever, suggesting that security measures can be minimized or removed. Its technologies also need to be supplemented with force, such that terrorists are hunted and killed. In this, a highly motivated and persistent offender is presupposed. Moreover, manifestations of innovation and displacement are not difficult to find.

Not long after the events of September 11, 2001, the US government crafted new policies and realigned government agencies (many within the newly organized Department of Homeland Security). Border security was transformed.[12] One such agency was the Transportation Security Administration (TSA), which now oversees the security of all airports within the United States.[13] TSA policies also affect security procedures at other airports throughout the world that connect passengers to the United States. In many respects, the TSA was rationalized on the grounds that it would safeguard the public through enhanced passenger screening at airports. Within two months, however, its enhanced security procedures were thwarted: Richard Reid attempted to detonate plastic explosives hidden in his shoes.[14] The TSA response was to require all passengers to remove their shoes during screening. In 2009, another bomb attack was thwarted when Umar Farouk Abdulmutallab failed to detonate the plastic explosives sewn into his underwear.[15] In response, the TSA introduced millimeter wave full-body scanners, which allow operators to see through clothing, observing nonmetallic objects.[16] Although one might assume that full-body scanners would extinguish any remaining security concerns, there is now speculation about surgically implanted explosive devices or "body cavity bombs."[17] In terms of displacement or substitution, recent years have witnessed the bombing of airports and subway stations, holiday resorts, tourist attractions, public events, and, among other sites, the office buildings of newspapers.[18] Terrorist attacks in the post-9/11 world have been scattered across the globe, but there is a unifying logic to them: they focus on governments that have provided discursive or material support to the US-led war on terror.

The contradictory construction of criminality—as an ever-present threat that can be regarded as neutralized if access to space is tightly regulated—leads security science to conceptualize crime control in territorial terms. In analyzing community policing in Los Angeles, Aaron Roussell shows how the LAPD conflates "community" with specific territories, or those geographical zones drawn up by the department and used to allocate resources. In adopting the "territorial imperative as organizational ideology," the LAPD construes policing as a matter of assigning officers to a section of the city and encouraging them to win and maintain control over their allotted territory.[19] For Roussell, this is problematic because how the LAPD imposes borders throughout the urban environment does not necessarily correspond to how residents conceive of community boundaries. Moreover, relying on territorial mappings devised by the LAPD is not geared toward empowering communities, but rather toward enhancing the geopolitical powers of police.[20]

In the context of security science, it could also be said that the wider public is not guaranteed to benefit from dependence upon territorial ideology and practice. Like urban policing, security science is primarily concerned with dominating particular territories rather than substantively addressing what it construes as a threat. Insofar as this is the case, it can rid itself of any concern with adjacent, proximal, or distant areas. Its fundamental mandate, after all, is to establish perimeters around specific spaces such that the entry of crime is barred. According to this logic, the exiling of criminalized behavior and threats to other territories is a matter for those entrusted with protecting such spaces.

The notion of territorialization is significant in a further, albeit distinct, sense. In chapter 1 we noted that "crime" is a label that may or may not be appended to behaviors.[21] As such, it can be understood to construct behavior in a very broad sense. Once behaviors are designated as "criminal," however, it is not uncommon to find that they are subsequently arranged or taxonomized. It has become "second nature" to arrange criminalized behaviors into a vertical hierarchy. At the top of this hierarchy, one finds those crimes deemed "serious," or felonious; at the bottom, one finds crimes deemed less serious, typically classified as misdemeanors. Not only, then, does "crime" impose a line that classifies behavior as legal or illegal; it also arranges the latter into a hierarchical order.[22] In the United States, the Federal Bureau of Investigation's (FBI's) uniform crime reports illustrate this tendency; murder, rape, robbery, aggravated assault, arson, burglary, and larceny are positioned as the most serious forms of offending.[23] Taxonomies and arrangements of crime do not simply correspond to some objective reality but are better understood as performative; they prescribe how specific behaviors ought to be understood and addressed. They also steer one's gaze in specific directions. Criminalized behaviors associated with the powerful, for example, are rarely central to hegemonic constructions of the "serious crime" category.

Security science intimates a departure from mappings of criminalized behavior that often pass for commonsense. In some important respects, it breaks apart and reterritorializes criminalized behaviors. Instead of arranging behaviors according to their "seriousness" or the harms they can be said to entail, and in lieu of preempting threats that can emerge, its construction and ordering of threats is informed by what its technologies observe and capture and the power asymmetries by which their operation is underpinned.

Insofar as this is the case, particular criminalized behaviors are pulled from their place in the "traditional" vertical hierarchy and relocated onto a horizontal plane. Security strategies promulgated under the threat of terrorism, for example, have morphed into an intensified form of policing air travelers for all kinds of "infractions": attempting to smuggle water onto a plane, exceeding permitted limits on the volume of gels and liquids, failing to place liquids in plastic bags of specified dimensions, or failing to remove laptops from carry-on luggage. The

TSA's airport security apparatus might prevent some acts of terrorism, but in practice, it spends most of its resources on confiscating water bottles, perfume, nail clippers and, alongside an array of other innocent objects, shampoo. Presumably the TSA's anti-terrorism measures originated with a vertical or hierarchical structure, used to arrange different crime types. Serious crimes were placed at the top, minor deviances toward the bottom. It would appear, however, that it did not take very long for innocuous conduct to be relocated to the horizontal plane that contains what most would recognize as more substantive threats.

The reterritorialization of criminalized behavior is also evident in the proliferation of CCTV and its associated private security apparatus. To accept the prima facie case that advocates of CCTV advance, the technology is intended to prevent serious property crimes and interpersonal crimes that involve the use of threats, force, and/or violence.[24] Much like the case of airport security, however, CCTV and private police inevitably end up devoting most of their resources to the surveillance of relatively innocuous behaviors because the ontology, epistemology, and technological capacities that constitute security draw these into the spotlight.[25] Homelessness, loitering, panhandling, and disorderly conduct emerge as appropriate objects for security technologies, thereby rationalizing approaches such as "broken windows" or "order maintenance policing."[26]

In addition, security science territorializes some practices as criminal or pulls everyday behaviors onto its plane of criminality. This tendency is illustrated, to give but one example, by the criminalization of jokes. As Lauren Martin points out, the TSA now broadcasts the following message at airports throughout the United States: "Making jokes or statements regarding bombs and/or threats during the screening process may be grounds for both civil and criminal penalties."[27] Given that numerous individuals have been arrested and prosecuted for bad jokes and sarcastic remarks, the TSA's message cannot be reduced to an idle threat. Martin is correct to interpret the prohibition on jokes as one manifestation of a broader strategy to produce disciplined, docile subjects who will not interfere with the imperatives of state security and anti-terrorism strategies. We discerned a comparable tendency in the area of bio-forensics, where DNA dragnets recrafted the exercise of legally protected rights into social deviance. In this instance, we see the construction of not only new deviances, but new classes of criminal behavior. To be sure, and akin to those taxonomies of crime that pass for common sense, this reterritorialization of behavior constitutes a political exercise in distilling signal from noise: everyday resistances (e.g., jokes, sarcasm) and behaviors associated with the powerless are criminalized and subject to control.

BUYING SECURITY

It is reasonable for governments to invest in the safety and well-being of their citizens. Indeed, that is a core justification for the existence of the state.[28] Spending $1 million on a levee that prevents $100 million in storm damage is prudent, and spending $10 million on a vaccine that prevents $1 billion in public health costs is wise. But spending $10 million on a sea wall to prevent $1 million in property damage is folly; the cure is worse than the disease. Judge Learned Hand's calculus of negligence provides one mechanism to decide what level of security investment is appropriate. Hand explained: "The owner's duty, as in other similar situations, to provide against resulting injuries is a function of three variables: (1) The probability that she will break away; (2) the gravity of the resulting injury, if she does; (3) the burden of adequate precautions. Possibly it serves to bring this notion into relief to state it in algebraic terms: if the probability be called *P*; the injury, *L*; and the burden, *B*; liability depends upon whether *B* is less than *L* multiplied by *P*: i.e., whether $B > PL$."[29]

In this formulation, Hand was describing the level of investment necessary to avoid legal liability for negligence, so it would be mistaken to assume that the state should invest in security only the product of possible injury multiplied by the probability of risk. But Hand's calculus of negligence can indicate whether or not investments in security science correspond to the threats they are designed to mitigate. This approach, for example, underpins William Landes's 1978 article on airport security and hijackings.[30] Landes determined that improvements in security mechanisms in US airports reduced the volume of aircraft hijackings, but did so at the cost of US$3.24–$9.25 million per hijacking.[31] In 2020 dollars, this represents $13.42–$38.31 million to deter a single hijacking. In a more innocent era, when hijacked airplanes were still principally used as getaway vehicles to flee to Cuba, not as weapons (to collapse skyscrapers filled with civilians and defense headquarters), Landes's analysis invited policy makers to ask if the wrongfulness and harm of one hijacking justified a multi-million-dollar price tag.[32]

Of course, in the immediate wake of 9/11, there was little concern for the price of security (social or fiscal). Patriotism swelled; jingoism bloomed. America became a *homeland*. First responders were venerated as heroes. Flags fluttered on every corner. In the face of such national trauma, any dispassionate Landes-style accounting would have appeared unpatriotic, even craven. Thus, it was a shock doctrine moment, an opportunity for a small group of actors to exploit a crisis in order to privatize public interests: "[Economist Milton] Friedman and his powerful followers had been perfecting this very strategy: waiting for a major crisis, then selling off pieces of the state to private players while citizens were still reeling from the shock, then quickly making the 'reforms' permanent."[33]

Instead of employing security as a collective social good, it is distributed according to market principles.[34] Security is monetized and taken to market; that portion of the population that can afford security consumes the product, while those who cannot afford it are segregated, marginalized, and disavowed, and profit is extracted in the process.[35] In this manner, stacks of public money are funneled into the coffers of the private security industry. Once rendered as a commodity within the marketplace, the revenues generated by private security firms appear unproblematic. However, these transfers of wealth reveal strong ties between privatized security provision and governments, and they are an important reminder that the public is both the consumer and the subject of security science.[36]

One week after the attacks, the US Congress granted the president broad authority to use military force in response to the terror attacks; then president George Bush announced the (still ongoing in 2021) war on terror.[37] And in the years that followed, the US government threw hundreds of billions of dollars at its pursuit of security. It bears noting that most of the actors in the arena of national security are motivated by profit over patriotism. Mike Battles, a defense contractor who secured $100 million in post-invasion Iraq contracts, put it bluntly: "For us, the fear and disorder offered real promise."[38] The world's two largest security companies—G4S and Securitas AB—have created empires out of security. G4S operates in more than ninety countries and employs approximately 570,000 people.[39] Over the 2010–2017 period, the company averaged a yearly revenue of £6.94 billion.[40] Securitas AB operates in more than fifty countries and employs upward of 340,000 people. In 2017, the company's revenue was around £7.7 billion. As documented by James Risen in *Pay Any Price*, the US Defense Department awarded more than US$400 billion to wartime contractors, many of whom have been sanctioned in fraud cases involving more than US$1 million.[41] That $400 billion in contractor costs—itself just a fraction of the cost of the war on terror—is a value seventeen times larger than NASA's 2020 operating budget.

Approximately US$10 million, for instance, was expropriated by eTreppid, a company founded by Dennis Montgomery and heavily funded by multimillionaire Warren Trepp.[42] Montgomery was a fraud who claimed that he had developed video-related computer technologies that (1) could compress and store large amounts of filmed footage, (2) offered unprecedented developments in the area of facial recognition software, and (3), his most extraordinary claim, were capable of identifying and decrypting messages supposedly concealed in al-Qaeda videos and Al Jazeera broadcasts.[43] His claims obviously appealed to government agencies that were seeking to identify and eliminate terrorists.

Montgomery convinced officials from the US Special Operations Command that his video technologies were genuine through rigged demonstrations. And as he wormed his way further into the war on terror, his supposed ability to decrypt secret messages within Al Jazeera broadcasts led to a number of US-bound flights being grounded. Several of these flights were operated by Air France. The French were unimpressed and demanded to examine Montgomery's decryption technology. It did not take long for the French to recognize Montgomery as a fraud, and that it was simply not plausible that secret messages were being encoded in al-Qaeda-produced videos or Al Jazeera broadcasts.[44] The officials who provided a no-bid contract to Montgomery were presumably not congenitally gullible, so exactly how did Montgomery successfully defraud the US Defense Department of US$10 million? Several factors made this possible. First, Special Operations Command was originally interested in eTreppid's video compression technology, not its more far-fetched al-Qaeda decryption technology. Second, because Montgomery involved co-conspirators (who pushed buttons when the software was supposed to detect a signal) in his fraudulent demonstrations, there was a plausible result, particularly because third, the US military, like most branches of government, is held in sway by an abiding fetish for technology. Fourth, the post-9/11 Defense Department had very deep pockets and could afford to invest in untested technologies. Fifth, and perhaps most importantly, it was imperative not that Special Operations Command actually catch Osama bin Laden and win the war on terror, but that they were *taking reasonable steps* to do so. It was important that they be neither actually negligent ($B > PL$) nor seen to be negligent. Just as there is a theater of security performed in airport passenger screenings, "providing the feeling of security . . . *instead of* the reality," there is also a theater of security at work within counterterrorism.[45]

The same logic explains the proliferation of surveillance. Research suggests that under specific conditions, CCTV can prevent particular forms of crime.[46] However, the overall crime-reduction benefits of CCTV are uncertain. It is not at all clear what role, if any, CCTV plays in reducing the volume of crime. Evidence is scattered and inconclusive, and the problem is compounded by the difficulty of subjecting CCTV to rigorous experimental logic.[47] To be sure, many scholars have expressed well-founded skepticism about the deterrent effect of CCTV.[48] "Statistical evaluations of the effect of videosurveillance on crime have shown that it is useless in some settings, yet cameras continue to be installed in those settings. . . . The logic of technology . . . is far stronger than negative findings."[49] It is perhaps telling that law enforcement experts concede that the value of CCTV resides in its retrospective capacities, helping to facilitate prosecution for acts that have been committed, rather than in preventing crime through the threat of possible punishment.[50] Nevertheless, throughout the 1990s and well into the twenty-first century, the UK government made massive investments in the installation of surveillance cameras in public spaces. Today, London, the CCTV capital of the world, is renowned for its "ring of steel."[51] Much of this money was distributed via a competitive bidding process in which business improvement districts (BIDs) and other stakeholders requested CCTV technologies. Exact figures are difficult to ascertain, but available literature suggests that between 1993 and 1999 the UK government subsidized CCTV to the tune of £243 million. In addition, the UK government had spent a further £500 million on CCTV surveillance in schools, hospitals, and transportation systems by 2005.[52]

B > PL.

CCTV—an inherently liminal technology—partitions "public space." The spaces that are subject to surveillance become "secured territories," while those that lie beyond are terra incognita, no-man's land. In his searing essay, "Beyond Blade Runner: Urban Control," Mike Davis explores the difference between secured and unsecured spaces.[53] By reimagining Ernest Burgess's iconic half-moon/dartboard representation of the American city (said to be "the most famous diagram in the history of social science"), Davis creates a new schematic for a dystopian Los Angeles:

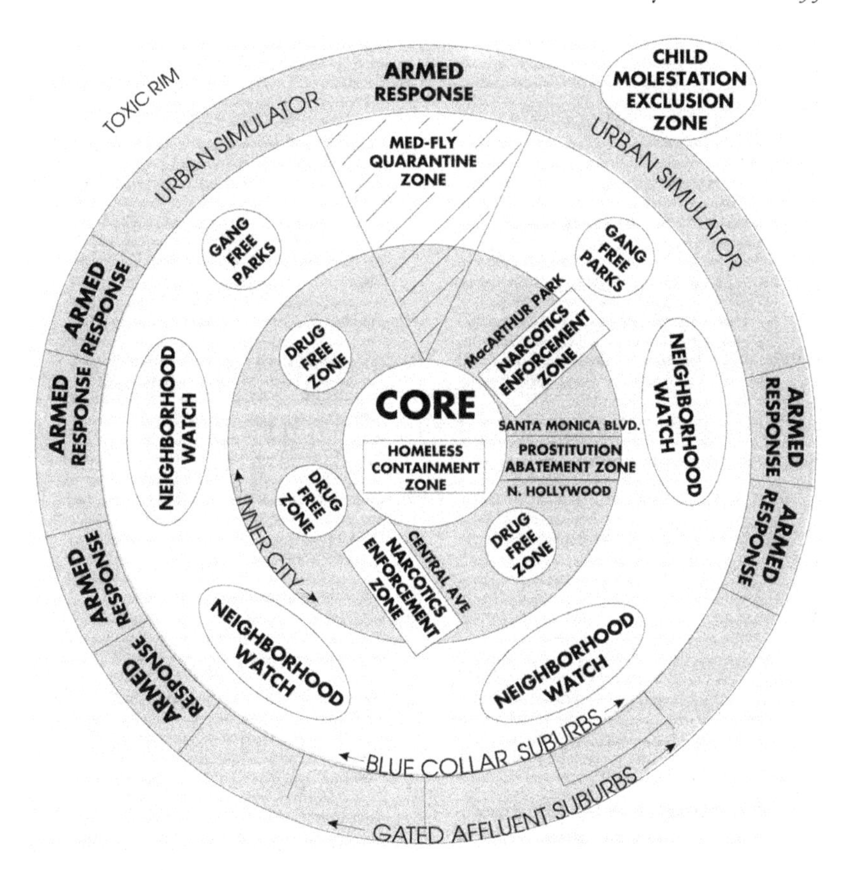

FIGURE 6. Mike Davis's "dystopian Los Angeles." *Source:* "Beyond Blade Runner: Urban Control (2): The Ecology of Fear," *Mediamatic* 8, nos. 2/3, January 1, 1995. Image reproduced with permission of Mike Davis.

> For those unfamiliar with the legacy of the Chicago School of Sociology and their canonical study of the *North American city* let me just say that Burgess' dart board represents the five concentric zones into which the struggle for the survival of the fittest (as imagined by Social Darwinists) supposedly sorts urban social classes and housing types. It portrays a "human ecology" organized by biological forces of invasion, competition, succession and symbiosis. My remapping of the urban structure takes Burgess back to the future. It preserves such "ecological" determinants as income, land value, class and race, but adds a decisive new factor: fear.[54]

At the center of Davis's image is downtown, where a *scanscape* of CCTV and networked systems allows white-collar employees and middle-class tourists to feel safe. "Inevitably the workplace or shopping mall video camera will become linked with home security systems, personal 'panic buttons,' car alarms, cellular phones, and the like, in a seamless continuity of surveillance over daily routine."[55] These technologies create a façade of security. Beyond the fortified core of the *scanscape*, however, lies the halo of barrios and ghettos that Davis denotes the "free fire zone." It is here, beyond CCTV's unblinking gaze, that undeterred social problems can be displaced. Insofar as the default logic of CCTV involves constructing secure perimeters for those who can afford them, the displacement of undesirable behaviors to areas beyond circumscribed spaces is functionally indistinguishable from deterrence.[56] Homelessness only becomes a social problem when it interferes with the flow of capital in the downtown region. It is the *visibility* of homelessness that renders it problematic.[57] In the outer *malebolgia* of Los Angeles, homelessness finds its ecological niche, blending in alongside poverty, racism, violence, desperation, and fear.

A FETISH FOR TECHNOLOGY

Many factors enabled Dennis Montgomery to defraud the US government of $10 million. But one of those factors, and one with great theoretical significance, was that the Special Operations Command—like most of government—was credulous about the promise of technology. David Harvey characterizes society's approach to technology as fetishistic: "endowing real or imagined objects or entities with self-contained, mysterious, and even magical powers to move and shape the world in distinctive ways."[58] Although modern technology is very real—tangible, visible, and commodified—it also appears as another form of magic. Hence Arthur C. Clarke's Third Law: "Any sufficiently advanced technology is indistinguishable from magic."[59] And he had a point: the technologies that are entirely quotidian today would have been miraculous a century ago:

> Although our jaded eyes no longer see the miracles in our everyday activities, we live in homes that are heated and cooled with central air, that are wired for electricity and natural gas and cable TV, and that have running hot and cold water and garbage disposals and toilets; we prepare our meals in microwave ovens; we watch DVD (or HD-DVD or Blu-ray) movies on high-definition televisions. We call our friends on cell phones, and check our email on Blackberries, and use the Internet to make instantaneous purchases from halfway around the world. We take photos on digital cameras and edit home movies on laptop computers. In vast cities with populations in the millions,

> we drive to work in automobiles that are capable of 100+ mph speeds, and when we arrive at our offices, we use desktop computers that possess five to ten times more computational power than the system that put man on the moon. We pick up food from drive-up windows, pick up our children from public schools, and shop in malls that contain dozens or even hundreds of stores. We get annual flu shots, take Viagra for sexual dysfunction, Prozac for depression, get LASIK for perfect vision, and get liposuction for perfect swimsuit bodies. We work out in gyms, each of us wrapped in the cocoon of an iPod, and grumble about being treated like cattle when we fly coach class between the United States and Europe. These things have become so ingrained in our lives that we no longer recognize them as extraordinary.[60]

Advances in transportation, communications, warfare, and computing power have all witnessed exponential development across human history. Graphed over time, the advances that have been realized in the last century would leap off the page.[61] The internet has reshaped the world's economies (the five largest corporations in the world—Microsoft, Apple, Amazon, Alphabet, and Facebook—are all digital companies) and transformed society. Of course people expect technology to move and shape the world in seemingly magical ways; it does so every day.

But while technology opens up new avenues of possibility, the social consequences of technology are underdetermined. This is not to say that all technologies are value neutral. Using Harvey's examples, it is very easy to see the terrific democratic and decentralizing potential of the printing press and the internet, and much harder to see it in the case of nuclear power.[62] But it does mean that power asymmetries—not the technology—determine how technology is employed and what it ultimately means. This point is essential in understanding that although technological innovations have the potential to *transform* power relations, they are more often used to *maintain* them: "All that we need to do is to reconstruct our social relations and our *mentalités* accordingly and we will enter into a brave new world. Yet immigration lines at borders get longer, protective walls increasingly surround whole states as well as gated communities, and the uses of space and time in daily life become ever more subject to the imperatives of capital circulation (speed-up and perpetual time-space compression). All this is backed by legal and police powers, most particularly to protect the core of private property rights in the maelstrom of change."[63]

The USA Patriot (Providing Appropriate Tools Required to Intercept and Obstruct Terrorism") Act was massive—more than three hundred pages long—and was passed with broad bipartisan support just forty-five days after the 9/11 attacks.[64] Many of the legislators who voted for it had not read the bill.[65] The PATRIOT Act was sweeping in scope, with titles to, inter alia, improve border security, enhance domestic security against terrorism, increase information sharing between law enforcement agencies and intelligence agencies (dissolving firewalls and blurring the lines between domestic law enforcement and national security), and remove obstacles to investigating terrorism (including "sneak and peek" searches, "business records" requests, and national security letters [NSLs], authorizing FBI agents to obtain records and data without probable cause or judicial warrant, and gag orders for those who have received an NSL). The PATRIOT Act was an unprecedented piece of legislation, couched in terms of security, but in practice it has been used for traditional law enforcement—for securing not the national interest as such, but private property rights. Between 2003 and 2005, the FBI issued more than 140,000 NSLs, but only 53 led to referrals to prosecutors: 19 were for fraud, 17 were for money laundering, and 17 were for immigration violations. None were for terrorism.[66]

In 2013, documents released by Edward Snowden revealed an expansive global surveillance program (known as PRISM), in which the National Security Agency (NSA) harvested millions of email messages of ordinary Americans.[67] Surveillance programs such as ECHELON enable the NSA to capture almost all electronic communications transmitted anywhere in the world.[68] Accumulating the data is within the technical capacity of the NSA and other government agencies, but comes at great cost. Shayana Kadidal estimates that it costs approximately $140 million to store the phone traffic that passes in and out of the United States over the course of one year. That is an immense sum of money, although well within the NSA's estimated annual budget of ~US$12 billion.[69] The ultimate purpose of accumulating such massive data is not clear. One can only imagine how much "noise" the NSA must store as it retains phone metadata, internet browsing histories, email content, financial records, and other data. Some have speculated that the primary intention is to test the feasibility of developing algorithms and risk prediction instruments to preempt terrorist attacks.[70] In any case, data accumulation of this kind has by no means ceased, despite revelations of its existence. If anything, it has become standard operating procedure in the United States (and elsewhere). The remit of the global intelligence community is no longer limited to neutralizing the threat of terror. It now seeks to "Collect it All," "Process it All," "Exploit it All," "Sniff it

FIGURE 7. DARPA's total information awareness logo. *Sources:* Information Awareness Office, home page; and "Information Awareness Office Seal," *Wikipedia*. n.d.

All," and "Know it All."[71] The community seeks to gather every thread of data that radiates from every person:

> As every man goes through life he fills in a number of forms for the record, each containing a number of questions. . . . There are thus hundreds of little threads radiating from every man, millions of threads in all. If these threads were suddenly to become visible, the whole sky would look like a spider's web, and if they materialized as rubber bands, buses, trams and even people would all lose the ability to move, and the wind would be unable to carry torn-up newspapers or autumn leaves along the streets of the city. They are not visible, they are not material, but every man is constantly aware of their existence.[72]

The all-seeing eye in the Defense Advanced Research Projects Agency (DARPA) Total Information Awareness logo displays this ambition.

Although it is easy to project the invasiveness of the PATRIOT Act and DARPA's all-seeing "Eye of Providence" onto a totalitarian canvas, transposing it for an Orwellian dystopia of telescreens and torture rooms, that might be a mistake.[73] The future of total information awareness might be far more subtle and far more insidious. In early 2019, the Chinese government developed a mobile app (Map of Deadbeat Debtors) that allows users to identify people who owe money within five hundred meters of the user. If it appears that identified individuals possess the means to repay their debts but are not doing so, they can be shamed or reported. This app is part of a much larger Chinese social credit score system, scheduled to be fully implemented for 1.4 billion citizens in 2020 but still under development in 2021, in which those who engage in antisocial activities (like playing video games or obstructing footpaths) can be barred from travel, getting loans, or obtaining employment.[74] Coercive collectivism, rather than some dystopian vision of "a boot stamping on a human face—forever," might signify the surveillance state's alternative potential.[75]

The specifics of the technology, and the specific ends to which that technology are enlisted, vary enormously, but across different situations what is unifying about technology fetishism is its faith in technology's capacity to remedy every social problem, that technology is viewed as inevitable and unstoppable, that it is seen as good (or sometimes as bad), and that it shapes and determines social changes (not the other way around).[76] Technology is agentic. This faith in technology—the belief that it creates our social arrangements and not that social relations create technology—absolves people from the consequences of their decisions. That technology will deliver us from evil enables policy makers to ignore the power asymmetries that contextualize what are perceived to be problems and to focus, more modestly, on their management using the best available technologies.[77]

That the problems identified by security science remain despite efforts to manage them is neatly illustrated with the example of commercial airline security.[78] After Landes demonstrated that airport security measures reduced aircraft hijackings, subsequent researchers noted that enhancing security could indeed reduce particular, targeted behaviors, but that these interventions often produced "displacements," increases in other undesirable behaviors, and analogous "substitution" effects.[79] Rather than being actors whose inclinations could simply be suppressed (the criminal man, just rational enough to be deterred, assumed by security science), potential hijackers appear to have innovated. Jon Cauley and Erik Iksoon Im concluded that "the installation of metal detectors [at US airports] was effective in both the short and long run [in deterring aircraft hijackings]. However, this policy was accompanied by a significant substitution effect that offsets, to some extent, the beneficial deterrence derived."[80] Walter Enders and Todd Sandler noted the same phenomenon: "In the case of metal detectors, kidnappings increased; in the case of embassy fortification, assassinations became more frequent."[81]

More than twenty-five years have passed since Cauley and Im noted that "substitution effects" often render new policies ineffective in "curbing terrorism per se."[82] But if anything, the national trauma of 9/11 only exacerbated faith in technology. As we noted earlier, since the establishment of the TSA, its developments in security measures have been spawned and driven by shifts in terrorism practices: Reid's shoe bomb is addressed by having all passengers remove their shoes at airport screening points; Abdulmutallab conceals explosives in his underwear, and the full-body scanner is installed.[83] Alongside these failures to suppress via perimeters and tightly regulated entry points—in short, spatial controls—it is difficult to discount the narrative that terrorists have redirected their attention, evidenced by bombing subway stations, holiday and tourist destinations, large public events and gatherings, and so on.[84]

There is an important lesson for security science in this. Its technologies presume that controlling the spatial correlates of criminalized practices is sufficient and therefore ought to be prioritized. But the terrorist subject is not limited to means and tactics. Those who engage in terror often engage in sophisticated reasoning about value and meaning, justice and purpose. It is not a coincidence that most of the terrorist acts that preoccupy the Anglophone world originate in areas that have been economically and politically devastated, to some significant degree, by the foreign policies of global superpowers. It is foolhardy to think that powerful governments like those of the United States, the United Kingdom, and Australia can sponsor foreign policies that engender widescale misery without simultaneously breeding discontent, hostility, and resentment. One form of terrorism inspires another. The real question should be how can the West *not* expect that some portion of extremely disadvantaged populations will come to embrace terrorism? How can one sustain the narrative that security science is a sufficient strategy for barring terrorism from finding expression? Shortly before he was executed for the Oklahoma City bombing, Timothy McVeigh wrote: "Bombing the Murrah Federal Building was morally and strategically equivalent to the U.S. hitting a government building in Serbia, Iraq, or other nations. . . . Based on observations of the policies of my own government, I viewed this action as an acceptable option. From this perspective, what occurred in Oklahoma City was no different than what Americans rain on the heads of others all the time. . . . Many foreign nations and peoples hate Americans for the very reasons most Americans loathe me. Think about that."[85]

The types of perimeter construction that continue to be fetishized within security science cannot contain the resentment that animates terrorism. When the problem is configured this way, the inevitable conclusion is that those interested in solutions need to focus on the broader cultural frameworks, geopolitical conditions, and power inequalities in which terrorism is embedded.[86]

THE STATE AS A TWO-FACED GOD

Many critical scholars of security science have addressed the tension between security and individual liberties, focusing especially on rights to privacy, freedom of expression, and protection from unwarranted state interference. For example, several have noted that the proliferation of CCTV violates the reasonable expectation to privacy that people retain in their public lives.[87] Others have observed that using CCTV footage in "crime stoppers" programs undermines the doctrine of "innocent until proven guilty."[88] And in terms of private security forces, other scholars have noted that these technologies pave the way for corporate entities to usurp the authority of the government's policing function, and in doing so, jeopardize individual rights and liberties.[89] Others have challenged post-9/11 expansion of surveillance and control as an effacement of the separation of powers and constitutional guarantees to privacy, free expression and association, and freedom from unreasonable searches and seizures.[90] In this critical tradition, the state is understood as an entity that protects individual rights through formal laws, and the technologies of security are understood as threats to the legal protections afforded by the state. The remedy, therefore, lies in fortifying the formal protections and guarantees promised by the state. It has been argued that the state should do more to regulate the uses of CCTV; revoke the legislation that authorizes NSA spy programs; maintain, or perhaps fortify, the separation of powers doctrine; and develop mechanisms to ensure public oversight over security science.[91]

There is enormous value in these lines of argument. The tension between freedom and security is real, with profound implications for human well-being.[92] After all, any state that does not formally protect individual rights and liberties can more easily slide into authoritarianism. However, critics who fetishize formal protections do not recognize that de jure guarantees cannot ensure that actual infringements upon civil liberties will cease. A legal prohibition against police brutality did not protect Rodney King from a rain of baton strikes; moreover, his beating did not constitute brutality—a Simi Valley jury acquitted the four LAPD defendants. Thus, it can be a mistake to assume that nominal, legal protections necessarily translate into substantive practice.

The modern state is a Janus-faced entity. One face represents the formal state's legal protections, which articulate a guarantee of liberty and—under most circumstances—freedom from governmental violence. The second face, however, represents a "subterranean" state, one that regularly infringes upon the liberties of its citizens and exercises instrumental violence, directly and indirectly, whenever doing so appears to be in its interests. In analyzing the modalities of power that characterize modernity, Michel Foucault suggests: "The real, corporal disciplines constituted the foundation of the formal, juridical liberties. The contract may have been regarded as the ideal foundation of law and political power; panopticism constituted the technique, universally widespread, of coercion. It continued to work in depth on the juridical structures of society, in order to make the effective mechanisms of power function in opposition to the formal framework that it had acquired. The 'Enlightenment,' which discovered the liberties, also invented the disciplines."[93] And in a somewhat less dense formula: "[Although] the universal juridicism of modern society seems to fix limits on the exercise of power, its universally widespread panopticism enables it to operate, *on the underside of the law*, a machinery that is both immense and minute, which supports, reinforces, multiplies the asymmetry of power and undermines the limits that are traced around the law."[94]

There are not two actors—a state that confers liberty and a security sector that undermines that liberty; there is only the state, and its exercise of power on the underside of law is not "outside" or "beyond" the state but central to its functioning.[95] According to Foucault, the modern state actively appropriates innovations in the exercise of power insofar as it encounters problems of social order, or new crises in the management of populations.[96] As such, it is not the case that knowledges and practices arise, then proceed to threaten the state. Rather, it is the state's need to manage populations that summons new knowledge formations and authorized practices, even if such practices run counter to the state's promissory notes surrounding individual liberty. There are two principal mechanisms through which the state circumvents the formal protections of law: exceptions within the written law and the substitution of national security for criminal jurisprudence.

First, as codified, the written law is riddled with jurisprudential exceptions that have accumulated over time. The Fourth Amendment of the US Constitution provides one useful example.[97] On its face, the text of the amendment appears unambiguous: "The right of the people to be secure in their persons, houses, papers, and effects, against unreasonable searches and seizures, shall not be violated, and no Warrants shall issue, but upon probable cause, supported by Oath or affirmation, and particularly describing the place to be searched, and the persons or things to be seized." And given the colonial Framers' great antipathy for the indiscriminate searches that were conducted under the British auspices of "general warrants" and "writs of assistance," it is only natural that "searches conducted outside the judicial process . . . are per se unreasonable under the Fourth Amendment—subject only to a few specifically established and well-delineated exceptions."[98] Those exceptions, however, are neither few nor well delineated. The exception has swallowed the rule. One article published in 1985 cataloged more than twenty categories of exception:

> searches incident to arrest . . . automobile searches . . . border searches . . . searches near the border . . . administrative searches . . . administrative searches of regulated businesses . . . stop and frisk . . . plain view, open field seizures and prison "shakedowns" . . . exigent circumstances . . . search of a person in custody . . . search incident to nonarrest when there is probable cause to arrest . . . fire investigations . . . warrantless entry following arrest elsewhere . . . boat boarding for document checks . . . consent searches . . . welfare searches . . . inventory searches . . . driver's license and vehicle registration checks . . . airport searches . . . searches at courthouse doors . . . the new "school search" . . . and finally the standing doctrine.[99]

Since 1985, several additional exceptions have been added to the jurisprudence of warrantless searches, including FBI NSLs (which, as described previously, even preclude the party searched from revealing that a search has occurred). In cases that do not fall into an exception, however, a violation of Fourth Amendment rights triggers the exclusionary rule—unless one of the enumerated exceptions to *that* rule should be satisfied.[100] Of course, it is not only case law that carves out interstitial exceptions within the formal laws of the state. A range of institutionalized practices runs counter to formal protections. These include forms of corruption such as

perjury,[101]

planting evidence,[102]

taking bribes,[103]

protection rackets,[104] and

coercing false confessions,[105]

as well as other tactics such as

using excessive force in policing,[106]

obtaining "consent" to engage in searches when probable cause is lacking,[107]

exercising discretion in discriminatory fashion,[108] and

overcharging criminal defendants in order to coerce guilty pleas.[109]

Today, in practical terms, prosecutors—not judges—decide case outcomes.[110] And because defendants pay a "trial penalty" for exercising their Sixth Amendment rights, the jury trial is in danger of disappearing altogether in the United States.[111] All of these developments undermine constitutional rights, putting any claim to the formal protections of law further out of reach for criminal defendants.

The second manner in which the state can circumvent the formal protections of law is by shifting the nature of discourse, from a jurisprudential one of criminal law and criminal procedure to one of national security. Criminal law and national security are not discrete concepts, of course; terrorism is a federal crime.[112] But passage of the Authorization for Use of Military Force in 2001 militarized the US response to terror and thereby placed national security concerns upon a different, and far less juridical, plane. Improperly obtained information might not be useful as evidence, but that does not mean it lacks national security value as intelligence.[113] Here, comparing the pre-9/11 case of Timothy McVeigh with the post-9/11 case of Zacarias Moussaoui is illuminating.

The federal prosecution of McVeigh for the bombing of the Alfred P. Murrah Federal Building in Oklahoma City was high profile, but in form resembled any traditional murder trial. McVeigh directed his lawyer to employ a necessity defense, but his lawyer demurred and instead adopted an unsuccessful strategy of mitigation.[114] In June 1997, McVeigh was found guilty on all eleven counts; he was sentenced to death in August 1997, and in 1999 and 2000, his respective appeals to the tenth circuit and the US Supreme Court were rejected. But just six days before McVeigh's execution, the Department of Justice "found" three thousand pages of potentially exculpatory evidence (which, under the *Brady* rule, should have been disclosed to the defense before trial), thereby prompting Attorney General John Ashcroft's decision to delay the execution by one month. When McVeigh's request for a stay of execution was rejected, he waived any further appeals, and he was executed by lethal injection on June 11, 2001 (three months to the day before four hijacked aircraft would strike the World Trade Center and the Pentagon).

In contrast, the federal prosecution of Zacarias Moussaoui shifted the nature of terrorism trials from traditional criminal practice in the direction of national security. One former prosecutor describes the Moussaoui case as "a sad relic of the past, a modern version of *Bleak House*, a bitter reminder of a time when we naively believed that terrorism was more a law enforcement problem than a national security problem."[115] Moussaoui, a French citizen and an al-Qaeda operative, was sometimes identified as the "twentieth hijacker," although he did not participate in—and was never demonstrably tied to—the 9/11 attacks. He was therefore charged with a number of conspiracy crimes: international terrorism, air piracy, weapons of mass destruction, and so forth. Moussaoui was permitted to represent himself at trial, but when he sought to subpoena classified documents and to call as witnesses imprisoned members of al-Qaeda, the prosecution objected, claiming that to do so would compromise national security. But Supreme Court precedent is clear, noting that it would be "unconscionable" to invoke government privilege to deny a defendant materials that could be used in a defense.[116] The Moussaoui case was also plagued by witness tampering—the TSA attorney contacted seven witnesses in breach of a sequestration order—which "nearly derailed the Moussaoui trial."[117] Before the case went to a jury and before Moussaoui was spared death and sentenced to six consecutive life terms, the presiding judge wrestled with blurred lines between criminal prosecution and national security, clearly troubled by the appearance of due process violations in a death penalty case.[118] Judge Leonie Brinkema wrote, "To the extent that the United States seeks a categorical, 'wartime' exception to the Sixth Amendment, it should reconsider whether the civilian criminal courts are the appropriate forum in which to prosecute alleged terrorists captured in the context of an ongoing war."[119]

"Reconsider" is exactly what the US government did. Instead of trying terror suspects as criminal defendants in US federal courts, many were detained indefinitely and interrogated as enemy combatants while in the custody of the US military in Guantánamo Bay, Cuba. Some were tortured, psychologically and physically.[120] Although the United States had prosecuted Japanese military officials for waterboarding as a war crime during the International Military Tribunal for the Far East (IMTFE), legal opinions suggested that "enhanced interrogation" practices, including waterboarding, did not satisfy the definition of torture.[121] Guantánamo detainees were neither afforded the Geneva Convention's protections of prisoners of war nor entitled to challenge their detention in US courts. Legal commentators have described Guantánamo as a "place outside the law," "a prison outside the law," and a "legal black hole."[122] This characterization, however, is not entirely accurate.[123] In a series of landmark cases, the US Supreme Court held that Guantánamo detainees possessed some basic rights.[124] Within Guantánamo itself, quasi-judicial proceedings called combatant status review tribunals (CSRTs) were organized to ascertain whether detainees met the definition of enemy combatant. These proceedings were controversial, and they differed in a number of essential respects from US criminal trials: "The CSRT applies a broad definition of 'enemy combatant' that inevitably ensnares innocent people; applies a presumption of guilt; has no juries; disables prisoners from gathering exculpatory evidence; prohibits prisoners from having lawyers; and relies on hearsay, coerced confessions, and secret evidence to reach its judgment that a prisoner should be detained indefinitely."[125]

Analogous military tribunals were constituted to try enemy combatants and to sentence them to imprisonment or death. Their legitimacy was rejected by many detainees. Addressing the nominal change in administrations, one mockingly noted, "Same circus, different clown."[126] In 2006, the US Supreme Court rejected these tribunals as unconstitutional, and Congress responded by passing the Military Commissions Act of 2006 (which, among other things, barred enemy combatants from lodging habeas corpus claims with US courts).[127] But in 2008 the Supreme Court parried, holding that the Military Commissions Act unconstitutionally suspended the right to habeas corpus promised in Article I of the Constitution.[128] This litigation was playing out almost seven years after detainees were taken into US custody. There are *still* detainees in Guantánamo Bay, now thirteen years after Barack Obama's vow in 2008 to shutter the notorious "prison outside the law." At the time of writing this book, a 2021 military commission trial has been scheduled for Khalid Sheikh Mohammed and four others, all facing the death penalty for plotting the 9/11 attacks.[129]

Although the military commission proceedings might very well be "rigged," the 9/11 defendants will be afforded a trial of a sort.[130] The same cannot be said for Anwar Nasser al-Awlaki. Al-Awlaki was a Yemeni American cleric, born in the United States, raised in Yemen, and university educated in the United States. He served as imam for a mosque in northern Virginia and as Muslim chaplain at George Washington University. After 9/11, he was identified as a moderate who could help bridge the gulf between East and West. But soon after he was charged with passport fraud (listing his place of birth as Yemen instead of New Mexico), al-Awlaki was included on US lists of terror suspects, prompting him to relocate to the United Kingdom and then to Yemen. In Yemen his views became increasingly militant, and he was imprisoned for eighteen months during 2006 and 2007 for involvement in an al-Qaeda conspiracy. Sometime in 2006, al-Awlaki began broadcasting radical online lectures, many on the topic of jihad, which circulated in the United Kingdom and elsewhere. Intelligent, charismatic, and a native English speaker, al-Awlaki was an influential spokesperson for radical Islam. One commentator described him as "the most dangerous ideologue in the world."[131] Although Presidential Executive Order 11905 prohibits assassination for political reasons, in April 2010, President Barack Obama authorized the extrajudicial killing of Anwar al-Awlaki as an imminent threat.[132] In this sense, he was designated as *homo sacer*, the accursed man outside the law who can be killed with impunity.[133]

Four months later, al-Awlaki's father, with assistance from the American Civil Liberties Union (ACLU) and the Center for Constitutional Rights, filed a remarkable lawsuit seeking injunctive relief. He asked the US government not to kill his son, a US citizen, "unless he presents a concrete, specific, and imminent threat to life or physical safety, and there are no means other than lethal force that could reasonably be employed to neutralize the threat" and he also asked the government to disclose the criteria by which US citizens are added to the "kill list." In December 2010, Judge John D. Bates's eighty-three-page decision artfully ducked the core legal question (whether the president can authorize the assassination of a US citizen), concluding instead that al-Awlaki's father lacked the legal standing to litigate the issue (as well as dismissing the suit on the basis of the political question doctrine, and without addressing the government's assertion of the state secrets privilege).[134]

Thus, although he had never been charged by the United States with any terror-related crime, Anwar al-Awlaki was killed in Yemen on September 30, 2011, when a US drone strike launched an AGM-114 Hellfire missile at his vehicle. Another US citizen, Samir Khan, was also incinerated in the strike. Two weeks later, another US drone strike killed al-Awlaki's sixteen-year-old son, Abdulrahman, also a US citizen, and in 2017, US SEAL Team 6 shot and killed al-Awlaki's eight-year-old daughter, Nawar.[135]

It is difficult to reconcile the extrajudicial killing of al-Awlaki with the language exhorting the rule of law in the Supreme Court's decision in *Boumediene*. Writing on behalf of the court, Justice Kennedy claimed, "The laws and Constitution are designed to survive, and remain in force, in extraordinary times. Liberty and security can be reconciled; and in our system they are reconciled within the framework of the law."[136] The distinction, of course, lies in the decision to treat a subject as existing within a legal framework or as outside that frame. "Sovereign is he who decides on the exception."[137] Giorgio Agamben has described the state of exception that allowed the Nazi government to flourish: "The entire Third Reich can be considered a state of exception that lasted twelve years. In this sense, modern totalitarianism can be defined as the establishment, by means of the state of exception, of a legal civil war that allows for the physical elimination not only of political adversaries but of entire categories of citizens who for some reason cannot be integrated into the political system."[138]

Agamben and others see the US response to 9/11 as the formation of a permanent state of exception.[139] The claim that in the Taliban, al-Qaeda, and Islamic State of Iraq and the Levant the US was engaging a new kind of enemy was used to justify the circumvention of traditional law enforcement authority and the use of military force to engage in asymmetrical conflicts. It was used to justify previously impermissible responses (e.g., extraordinary rendition, torture). And it was used to justify abridgment or effacement of traditional constitutional rights and international standards. But the war on terror is a war without end. "Oceania has always been at war with Eastasia."[140] Whole groups of people exist in a state of exception. Guantánamo remains a zone of exception. In fact, virtually all aspects of government can be positioned within national security so as to justify the exception: "When fully deployed, the SBU [sensitive but unclassified] category effectively expands national security to include any kind of information whose release might be inconvenient to the execution of state policy. When combined with the mosaic theory of information risk, there is literally no aspect of governmental work that could not be conceptualized as an essential part of U.S. national security and thus seen as a threat if made public."[141] Instead of being seen as an interruption in traditional practice, the shift from terrorism as crime to terrorism as national security issue should be viewed as an enduring movement, one that allows the government to easily recast criminal proceedings as matters of national security, unimpeded by constitutional restrictions.

CHAPTER 5

Environmental Crime Science

Missing the Forest for the Acronyms

In the previous chapter, the desire to create boundaries and enclosed zones was posited as the core principle of security science. The focus of this chapter is on environmental crime science, a related discourse in which the manipulation of space is construed as central to crime control. It seeks to create spaces and environments that are imminently crime proof. That is, physical spaces and environments are thought to be capable of design and redesign, continual modification, emitting nudges and signals concerning acceptable and unacceptable behaviors, and so on, such that the enacting of criminalized behaviors is inhibited. The idea of spaces that are imminently crime (and/or "disorder") proof is well illustrated by park benches that are designed so that it is impossible to sleep on them or, somewhat similarly, the installation of concrete spikes that compel homeless people to "move on."

The fetish for reengineering the environment can be found in a variety of perspectives that have gained traction over the last fifty years or so. By no means exhaustive, the tendency to see spatial features and design as holding the technological key to crime's resolution is evident in routine activities theory (RAT), situational crime prevention (SCP), crime prevention through environmental design (CPTED), broken windows (BW) and/or order maintenance policing (OMP), and "hotspot" policing (which is heavily dependent upon "compare stats" or COMPSTAT).

There are points of tension across these perspectives, but a brief overview of their core claims reveals the common ground—an alleged environment-crime relationship—by which they are connected. In the initial articulation offered by Lawrence Cohen and Marcus Felson, routine activities suggested that major changes in the organization of everyday life could explain fluctuations in aggregate crime rates. More specifically, Cohen and Felson suggested that American society experienced significant changes during the 1960s and 1970s: participation in the labor force and higher education increased, people were required to travel greater distances between home and work, more automobiles and consumer commodities entered the marketplace, and many people engaged in leisure activities that took them farther away from their homes and neighborhoods.[1] These new everyday routines, so the theory goes, have enhanced opportunities for criminal offending. The more people are absent from their homes, the easier it becomes to commit burglaries; more time spent in public places allows for offender and victim to cross paths, increasing the risk of interpersonal offenses; and theft feeds upon the mass production and widespread availability of consumer commodities.

From general routines and aggregate crime rates, the proponents of RAT are quick to shift gears and focus on the ingredients necessary for an isolated crime event to materialize. The model they offer parallels the firefighter's "fire triangle," which notes that for a fire to burn, the convergence of three elements is required: heat, fuel, and oxygen. If one of those elements can be eliminated, the fire goes out. In like manner, RAT posits that "crime" (i.e., various property and interpersonal crimes) requires the copresence of a motivated offender and a victim, as well as the absence of guardianship. If one of those elements can be eliminated, there is no crime. It is this third element that locates routine activities squarely within environmental crime science: incredibly broad in meaning, *guardianship* can refer to police officers on foot patrol, but it also refers to "target hardening" devices, design features that allow for greater visibility or transparency, alarms, and among other things, anti-theft devices—all of which amount to modifications of space.

Situational crime prevention consists of four components, according to Ronald Clarke. Taking its "theoretical" inspiration from routine activities, it purports to be informed by rational-choice models of the subject. From here, it is said to rely on quasi-experimental "action research" methods and focus on developing opportunity reducing techniques. Situational interventions are then evaluated with a particular focus on the problem of "displacement."[2] Somewhat obviously, "opportunity reducing techniques"—various ways of manipulating the

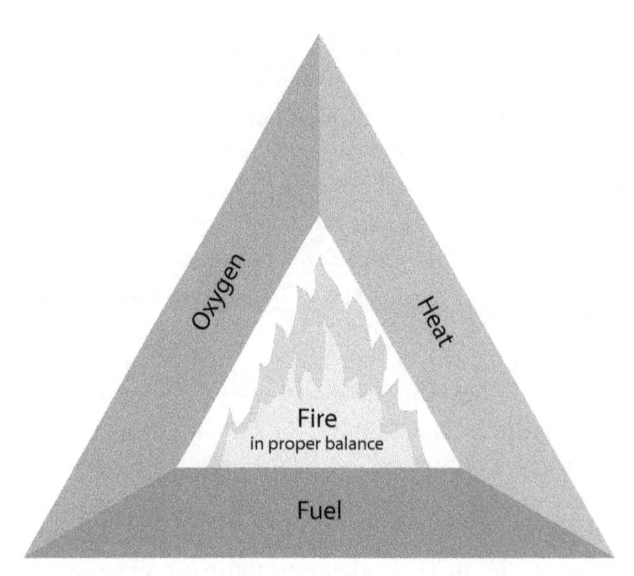

FIGURE 8. The fire triangle. Its simplicity does not seem to have prevented raging wildfires. *Source:* Wikimedia commons, Fire Triangle, n.d.

physical, technological, and social aspects of space—constitute the core of SCP. "Opportunities" for crime are held to reside within the environment, and these are exploited by "rational actors." Techniques and interventions focus on closing environmental loopholes that are conducive to particular crimes, and these techniques of closure are to be evaluated in terms of changes in crime rates.

As its very name suggests, CPTED is probably the most obvious discursive strand for the articulation and promotion of environmental crime science. It is more or less exclusively concerned with designing environments that are crime proof. In this sense, it elevates a handful of particular techniques that appear in other environmental perspectives to a position of prominence, especially strategies like "target hardening" (e.g., graffiti-proof surfaces, stronger fencing), increasing visibility and lines of sight for the sake of enhancing surveillance (e.g., street lighting, using glass or transparent building materials), and designing space such that the movement of people is modulated and controlled.[3]

The BW theory has received substantial public attention; as a result, its basic premises are well known. It suggests that if minor forms of physical and social disorder are left unchecked in a neighborhood, such as graffiti, public drunkenness, and loitering, then serious crime and urban decay will inevitably follow. According to proponents, it follows that controlling public behaviors and the physical appearance of the urban environment will lead to reductions in "serious crime," hence the strong push for graffiti removal; policing the homeless; and "crackdowns" on begging, public drunkenness, vagrancy, and other nuisance crimes.[4] In Rudolf Giuliani's New York City, there was a war on "squeegee men." These nuisance behaviors are understood as signal crimes, telling "serious offenders" that the community that tolerates them is not invested in order. Accordingly, the focus shifts to truncating minor crimes for what they supposedly communicate.

Broken windows provides the theoretical basis for OMP, which encourages officers to treat minor infractions and nuisance behaviors as grounds for making arrests or issuing tickets. Broken windows and OMP suggest that minor disorder is best controlled by ensuring that police officers remain a permanent fixture of the physical environment. Police resources should therefore be organized such that officers are available to walk beats and maintain a strong, visible presence within any given neighborhood. It is the emphasis on physical appearances and the public presence of police officers that locates the BW theory within the domain of environmental crime science.

"Hot spot" (or "problem-oriented") policing is similar to BW, but hinges on distinct conceptual underpinnings and, at least more recently, technological developments. It is premised upon the notion that crime is concentrated at particular times and places.[5] If data that record when and where crimes occur can be systematically gathered and quickly analyzed, patterns in offender behavior can be identified. Police officers can then saturate particular areas in which crime spikes have been discerned. While police departments have a long history of producing pin maps that plot crime patterns, things took a "technological turn" in the mid-1990s with the introduction of COMPSTAT. An abbreviation of "compare statistics," COMPSTAT is a computer program that facilitates the gathering, input, and analysis of crime-related information.[6] In the public imagination, it would seem as though these crime-mapping functions are regarded as the fundamental feature of COMPSTAT. Yet while it is used in this way, police departments that have embraced the technology typically utilize it to monitor and hold precinct commanders accountable for crime trends in their districts.

This chapter focuses on four lines of critique, drawing from different manifestations of environmental crime science for illustrative purposes. First, we suggest that environmental crime science parallels the fundamental motifs found in neoliberal rhetoric. Both discourses (a) construe the state as incompetent, such that private actors should be entrusted to control public life, (b) claim that privatization and the concentration of resources in the hands of powerful economic players will benefit everybody as positive effects "trickle down," and (c) responsibilize individuals with risk avoidance and management. This is not simply to suggest that environmental crime science mimics neoliberal rhetoric, but that in doing so, it pushes the logic of neoliberalism into the domain of "crime control."

Following this first line of critique, we explore the racism and classism that informs environmental crime science. This is probably not surprising given that neoliberalism is grounded in, and works to reproduce, racialized and class power asymmetries. Indeed, its rhetorical categories and defenses are inseparable from the racialized and class-based hierarchy that it instantiates. We extend the concern with racism and classism into an examination of some of the critiques that have been made of policing practices inspired by environmental crime science. Within available critiques, there is a tendency to recuperate environmental crime science by construing its theoretical positions as antithetical to the policing practices that it legitimizes. Apparently the theory and practice have somehow drifted apart. We argue that there is no such separation: the theoretical statements proffered by environmental crime science are racist and classist, and so too are the policing practices that it inspires. A second tendency in critique is to suggest that efforts to implement environmental crime science will undermine the perceived legitimacy of the state and its exercise of authority. Our argument here is that a legitimacy crisis should logically follow as a result of putting the principles of environmental crime science into practice. However, the state's perceived legitimacy is unlikely to be in peril, and as already noted, the racism and classism of the theory accounts for this. In other words, there is no legitimacy crisis, because environmental crime science overwhelmingly criminalizes society's most marginalized and powerless members.

Our third argument focuses on the politics of environmental crime science's construction of the offender. While proponents routinely posit that a rational choice model of the subject governs environmental approaches, this would not appear to be the construct that animates analysis. Rather, the offender is effectively understood in bio-ecological terms. On the one hand, the criminal actor may be an "opportunist," whose behaviors embody adaptations to situational contexts; on the other hand, the offender is construed as a "habitual criminal." Despite their irreconcilability, both constructs exonerate a politics of suppression by technological feats or incapacitation.

Finally, we suggest that the politics of environmental crime science are further deployed as a set of prescriptions for criminology. Accordingly, the criminologist is asked to valorize control and suppression, align their work and interests with disciplinary institutions and business interests, and focus on devising technological strategies that circumscribe possible courses of action. In this, social inequalities and deprivation are accorded ontological status but cast aside as irrelevant to criminology. As with other crime sciences, this amounts to a criminology that is complicit with contemporary power asymmetries and, if anything, works toward their preservation and entrenchment by inscribing criminality on the body.

RECRAFTING "CRIME CONTROL" ALONG NEOLIBERAL LINES

Throughout previous chapters, we have drawn attention to the politics of crime science, or how crime science and power asymmetries are interwoven. Discourses and practices that purport to control crime via environmental manipulations do not break with this pattern.

As a discourse, environmental crime science contains many moments that replicate the rhetorical strategies and categories that are used to defend neoliberalism. Here, we focus on three moments of overlap. First, both neoliberalism and environmental crime science routinely posit that the state is incompetent and inefficient. This argument is a mechanism to promote "deregulation" or, more precisely, to intensify the power of private interests to regulate and control the social body. Second, both discourses claim that their techniques and strategies, if adopted, will produce some greater social good. Whereas neoliberal rhetoric often encapsulates this sentiment in "trickle down economics," environmental crime science formulates it as "diffusion of benefits." Third, the two discourses are both wedded to the trope of "personal responsibility." This easily leads to "victim blaming" and, closely related to the mantra of "deregulation," absolving the state of any responsibility toward its citizens. As a corollary, structural arrangements and power asymmetries are obliterated from view, as the social is reduced to a series of discrete individual choices.

In our view, such discursive replication—state incompetence/inefficiency, privatization as delivering collective benefits, personal responsibility—partakes in recrafting the meaning of "crime control" such that it is forced into harmony with neoliberal logic and subjectivity.

The Incompetent State and Its Inefficiencies

That neoliberal rhetoric constructs the state as incompetent and inefficient is fairly well known. Construing the state in such a manner opens the way for neoliberalism to pursue the privatization and commodification of public assets or functions previously regarded as falling within the purview of states, such as utilities, education, health care, warfare, and criminal justice. As David Harvey points out, it is not that the state is entirely excluded as commodification processes unfold. However, the role of the state will be reconfigured as neoliberal promises of greater efficiency become hegemonic: rather than exerting control over economic interests, it is economic interests that will delimit the functional role of the state. Private interests will generally marshal the coercive capacity of the state to contain the social unrest that follows from their aggressive capturing of collectively produced wealth. In this, they secure profit while diminishing the general welfare and the capacity of the public to exercise autonomy and power in political affairs.[7] Furthermore, privatization erases the normative tone of services once deemed public and essential. Instead, public sector functions are treated like any other business, subject to managerial oversight and performance evaluations, and mined for profit maximization and revenue streams.

In one way or another, the various strands of environmental crime science incorporate the view that states cannot be entrusted with the task of controlling crime. Some accounts posit that nonstate actors are better positioned (and more motivated) than officials to devise and implement crime control strategies; in other accounts, environment-based logics intimate that state activity could be better rationalized if it were subject to managerial oversight or evaluated against performance expectations and measures.

The rhetoric of state inefficiency is especially evident in SCP's insistence that "standard" police tactics are limited. For police to be effective, they will need to find ways of being more innovative and creative. In their analysis of pay phone toll fraud, which involved "hustlers" stealing pin numbers from calling card companies and selling these to people wishing to make overseas calls, Gisela Bichler and Ronald Clarke note:

> Hustlers were not much deterred by the regular presence of large numbers of police. While the police were understandably frustrated by their inability

to deal with the problem through conventional means of surveillance and arrest, they missed the opportunity to take the lead in finding more creative solutions to the problem. . . . It was only when building managers decided that the problem must be solved were solutions identified and a spectacular crime prevention success achieved. . . . This is further evidence of just how much the police must change if they are to take advantage of the potential afforded by problem-oriented policing.[8]

Examining convenience store robberies in the US state of Florida, Ronald Hunter and Ray Jeffery find that there is a symbiotic relationship between environmental measures and business interests. As such, businesses are encouraged to take the lead in combatting crime problems that directly affect them. Doing so is held to be efficient and may have the additional benefit of preventing the state from interfering with business practices. Concerning the effectiveness of situational measures, they posit: "The results of Florida's robbery prevention efforts are being felt immediately. Convenience store robberies in Florida have declined from 5,548 in 1989 to 4,904 in 1990 despite an overall increase in violent crimes (from 145,473 to 160,544) during the same period."[9] It is curious that the 644 fewer store robberies is the noteworthy figure, rather than the 15,071 additional violent crimes. Paradoxically, the latter figure is used to highlight the effectiveness of environmental measures, even though it may just as easily be read as evidence of their limits. That environmental crime science can celebrate a decline in store robberies in a rising sea of violence without any sense of irony indicates that it is primarily invested in businesses and problems that impact profitability. There is, however, a further benefit for businesses that voluntarily take the lead in adopting situational prevention policies and measures: "Implementing them may also prevent mandatory legislation which might limit corporate discretion."[10] Not only can business be more efficient if it chooses its own modes of regulation; it can also act as a counter-sovereign to the state.

In terms of dissolving "crime control" in managerial logic, this is perhaps most evident in the rise of things like order maintenance and hot spot policing, and their association with COMPSTAT. As various scholars have noted, COMPSTAT's capacity to map crime problems in a precinct, fluctuations in the rate of particular crimes, and arrest rates has come to operate as a basis for new management strategies within many police departments, especially in the United States. COMPSTAT meetings, in which district commanders field questions about crime rates from police chiefs, are somewhat notorious. James Willis and colleagues note that such meetings are more "ceremonial" than substantive:

FIGURE 9. Mayor Bill de Blasio of New York City announcing COMPSTAT 2.0 in 2016. *Source:* Image courtesy of City of New York.

"District commanders were assessed primarily on their ability to provide responses to the chief's inquiries, not on the quality of that response. . . . Indeed, most of the middle managers focused on having facts and figures at their fingertips for the formal meeting, rather than finding ways to prevent and control crime in their districts on a day-to-day basis."[11]

Facts and figures concerning crime rates and police activity operate as "key performance indicators" (KPIs). Although such figures can be manipulated fairly easily, they are nevertheless reified, becoming more important than the substantive reality they purport to describe.[12]

Consistent with the neoliberal tendency to exclude the public from exercising control over public agencies, COMPSTAT's fetishization of figures offers a particular way of constructing effective policing. As Mark Moore concludes, COMPSTAT fosters the view among many police that the volume of arrests made embodies a good in itself. What falls out of view is the extent to which such arrests are consistent with principles of justice and represent the interests of the communities so policed.[13]

Trickle Down as Diffusion of Benefits

Neoliberalism neutralizes any concern raised by the specter of state incompetency by asserting that a capitalist economy, if left to its own devices, will work to the benefit of all. Adam Smith's "invisible hand" serves conveniently as capitalism's deus ex machina.[14] The excessive accumulation of resources by the few will inevitably be redistributed, and thus wealth will "trickle down" to the lowest rungs of the social ladder. Not surprisingly, critics of neoliberalism find no reason to accept the notion that fostering concentrations of wealth will somehow ensure the greater social good. Rather, securing the economic interests of a small minority entails suffering among the vast majority.[15]

In environmental crime science, the neoliberal mantra of "trickle down economics" surfaces in the notion of "diffusion of benefits." Insisting that controlling one particular type of offending via environmental modifications will be accompanied by additional benefits, even if these are unintentional or unforeseeable, acts as a rejoinder to skeptics and critics, who often emphasize the problem of "displacement."

According to the displacement thesis, environmental modifications are futile because they do not address the problem of motivation and dispositions toward criminality. As a result, controlling behaviors within a particular zone will simply lead to their relocation to areas that do not have situational controls in place. Alternatively, criminalized behaviors may be displaced in the sense that some offenders will shift practices or seek out innovations in offending. The emphasis on displacement often intersects with concern over social inequalities. Here, it is argued that environmental measures not only relocate crime, but that they push crime problems away from privileged areas and into spaces marked by social marginality.[16]

To be sure, whether environmental controls can be conflated with a "diffusion of benefits" or "displacement" is an incredibly limited and limiting debate. Both positions accord crime an ontological status, and both proceed to reduce crime to representational epistemology. That is, despite their irreconcilability on the surface, they operate by assuming that crime is an objectively given reality and that the task of crime control can be reduced to one of discovering what kinds of interventions have the most impact upon the volume of offending.

Even if one were to remain within the confines of representational epistemology, however, insisting upon a "diffusion of benefits" is peculiar given that it cannot be meaningfully demonstrated. In order to show that place-based strategies control some particular crime within an area and without displacement, one would need to "control" for an insurmountable range of possibilities: shifts to other forms of offending, behavior being relocated to other spaces (both physical and virtual), and adjustments to situational interventions over time. Within each of these broad possibilities, many varieties of adaptation are imaginable.

And yet environmental crime science is adamant that the notion of displacement is unfounded. Efforts to refute it can lead to arguments that necessitate bewildering contortions of logic. For example, Henda Hsu and David McDowall ask whether target hardening results in "deadlier terrorist attacks against protected targets?"[17] Formulating the problem in such a manner is antithetical to the very prescriptions of environmental logic: target-hardening strategies should not correlate with escalations in the behavior they target, nor should they be found to correlate with stability in targeted behaviors. The reason for posing the question in this way soon becomes obvious. Hsu and McDowall find that target-hardening efforts concerning aircrafts/airports and US embassies had no discernible effect upon hardened targets: "Overall, the analysis found no evidence that attacks with casualties changed in either frequency or proportion after adoption of the antiterrorist security measures. Casualty-producing strikes increased slightly for aviation targets and decreased slightly for diplomatic targets. These changes were small enough to be chance."[18]

The logic of environmental crime science dictates that situational measures constitute intervention points after which crime rates should decline. Hsu and McDowall, however, suddenly frame the lack of change or consistency in behavior as evidence that target hardening works because, it would seem, attacks against aircraft and embassies did not become any deadlier than before such measures were put in place. Moreover, they are compelled to acknowledge that their results cannot "speak directly" to the problem of "adaptation." Yet they reach the conclusion that "costly unintended consequences are not the likely end of situational prevention strategies," thus calling into question "long-standing fears of collateral damage surrounding the efficacy of target-hardening measures."[19] In no way has any such claim been substantively demonstrated. The only thing that has been shown is that target-hardening measures had no discernible effect. "Displacement" is discounted by equating it with the possibility of "escalation."

To give one further illustration of the will to negate displacement, in analyzing situational measures against check forgery, Johannes Knutsson and Eckart Kuhlhorn assert that displacement did not transpire after measures were put in place: "An examination of conceivable alternative crimes, e.g., burglary from homes, basements, and attics and thefts of and from cars, shows that they *declined* in the two years immediately following the implementation of the measures [taken to inhibit check forgery]."[20]

How burglary and thefts of and from cars constitute "conceivable alternatives" to check forgery is not entirely clear. The distinctiveness of the latter resides in its deceptive qualities and "indirect" victimization. A more appropriate set of alternatives would therefore be crimes of confidence, credit card fraud, identity theft, and so on. By Knutsson and Kuhlhorn's logic, proponents of environmental crime science may as well find any crime that shows a decline in prevalence and chalk this up as evidence that displacement could not be shown. To put this in terms often used by science and technology scholars, displacement is displaced by "rhetorical closure," a move that becomes necessary because the problem cannot be addressed from within the confines of environmental crime science's own epistemological standpoint.

That environmental crime science is indebted to the notion of "trickle down" is also evident in assuring that situational techniques of control, if implemented, can only be beneficial to the lives of those subject to them. Arguments along these lines illustrate the paternalistic and condescending nature of environmental crime science. According to Clarke, situational measures will inspire instrumentally rational individuals to lead a life that is "crime free": "Few offenders are so driven by need or desire that they have to maintain a certain level of offending whatever the cost. For many, the elimination of easy opportunities may actually encourage them to explore non-criminal alternatives."[21]

In examining state-orchestrated housing allowances and how they create space for welfare fraud (providing false information concerning income in order to receive an allowance), Kuhlhorn adopts an analogous stance, arguing that environmental controls will be appreciated by most insofar as they save individuals the trouble of grappling with their inner demons: "This redistribution function [of the welfare state] creates a tempting opportunity structure for white collar crime; a person who conceals his financial assets . . . can benefit by . . . cheating on housing allowances. . . . [On] the basis of a common income concept for sickness insurance and housing allowances, people submit a single statement and so avoid making mistakes or being subjected to too much pressure of temptation."[22] Aside from the not so subtle eliding of "white collar crime" and "welfare fraud," this passage is noteworthy for constructing individuals as weak willed and thus likely to succumb to temptations that arise from situational loopholes. It is almost as if individuals were akin to Eve and Adam being tempted by the serpent's apple. Situational measures, however, promise to unravel their dilemma by removing the tree of knowledge, thereby crafting a world that is beyond good and evil or, somewhat more mundanely, inhibitive of crime.

Repackaging the neoliberal category of "trickle down economics" as "diffusion of benefits" involves a contradiction of some import. On the one hand, environmental crime science promises to channel individuals along paths that it perceives as morally righteous: "noncriminal alternatives" will be pursued as "temptation" is nullified. On the other hand, this apparent choosing from among a range of possibilities occurs only in the absence of choice: environmental modifications promise to block "criminal" pathways. In this, environmental crime science posits that the eradication of choice constitutes the means by which moral action can be made to transpire; moral action is assured by technologies that suppress. The promise to ensure behavior deemed appropriate, then, would appear to be grounded in environmental crime science's own "amorality," a preference for exercising domination over the subject rather than according it autonomy.

Personal Responsibility and Victim Blaming

Like the rhetorical categories of state incompetence and trickle down economics, "personal responsibility" is by now a well-worn construct of the neoliberal era. Among other broad effects and implications, the notion divests the state of any responsibility toward its citizens and corrodes critical, sociological narratives that rely on concepts such as "power" and "structure" to explain individual and group trajectories. Whatever conditions and life experiences are found to accompany neoliberal orders, personal responsibility is deployed to deflect criticism. Individuals are held to make choices, and these explain poverty, suffering, and exclusion, but likewise account for wealth and privilege. Insofar as the state is absolved of duties toward its citizens, individuals are tasked with managing risks to their well-being.[23] If something goes wrong or if suffering is endured, this can be attributed to an error in judgment that the individual has made at some point in time.

Environmental crime science inscribes "personal responsibility" into its general conceptual architecture and, perhaps more obviously, into those moments when it provides guidance on how individuals can minimize their risks of victimization. Concerning the former, Clarke suggests that environmental approaches are simply a "scientific extension" of what common sense dictates: "Every day, we all do such things as lock our doors, secure our valuables, counsel our children, and guard our purses to reduce the risk of crime. . . . It is into this group of crime control measures that situational crime prevention fits. Indeed, it can be regarded as the scientific arm of routine precautions."[24]

The suggestion that environmental crime science does little more than scientize a range of behaviors and practices that people adopt as a matter of "routine precaution," if not instinct, is subsequently refracted in specific contexts. To insist that people can take precautionary steps that preempt, and thus inhibit, victimization, may not appear all that objectionable at first glance. However, it does not take much for the responsibilizing of individuals to slide into victim blaming.

Presuming that individuals can exercise control over their risk of victimization occasionally leads to advice that is muddled and perversely paradoxical, so much so that we can only hope no one ever tries to put it into practice. In concluding their analysis of carjacking, for example, Terance Miethe and William Sousa note: "Depending on the particular combination of other characteristics in the carjacking situation, victim resistance may be a successful intervention or the precipitating factor for serious injuries or death."[25] What, one might ask, are these "other characteristics in the carjacking situation" that should be considered in determining whether resistance is a wise strategy, bearing in mind that serious injuries or death may occur if resistance fails? The potential victim ought to assess the interplay of four other variables: (a) whether they are alone or with others; (b) the relative numbers of offenders and victims; (c) other personal valuables present; and (d) type of weapon, if any, possessed by the assailant.[26] If the human capacity for rational calculation were akin to that afforded by advanced statistical analysis programs, this inducement to "weigh up" the carjacking situation—paying particular attention to how outcomes are altered by throwing resistance into the equation—might make sense. We very much doubt, however, that a carjacking will bring out the "inner SPSS" in anyone.

While they flirt with the idea of victim resistance, often encouraging it, Miethe and Sousa do suggest that less heroic courses of action are available to individuals.[27] Their data reveal that many carjacking incidents occur in parking lots and begin "when victims [are] outside their vehicles." This leads to the suggestion that "the most efficient [SCP] strategy for these carjackings may simply involve increased public awareness" of "vulnerability in these particular settings."[28] One could be slightly bemused by this: an extensive body of data is gathered, numbers are crunched, statistical tables are produced, and the advice that is distilled through this lengthy process appears to be "remain alert, especially in carparks." Proffering this kind of advice makes it seem as though victimization could be minimized if only individuals were willing to make simple adjustments to their everyday behavior and recognize their vulnerability.

Asserting that one should "remain alert" may seem embarrassingly banal, but it is precisely this level of banality that harbors environmental crime science's proclivity for victim blaming: "Did *you* remain alert while returning to your car?" That environmental crime science does not hesitate to put such questions to victims intimates the privileged, white, masculinist logic by which it is interpellated.

David Finkelhor and Nancy Asdigian argue that responsibilizing individuals for crime avoidance is problematic because it fails to recognize that some characteristics associated with victimization are beyond one's control, such as gender, sexual orientation, ethnicity, and the (physical and mental) capacity to resist.[29] Hate crimes and men's sexed violence against women, for example, typically involve dominant groups targeting those from marginalized social locations or minority groups with little social power. Most victims of rape and sexual assault know the offender.[30] Suggesting that such behaviors can be addressed by encouraging individuals to take precautionary steps minimizes the idea that more attention ought to be paid to those who engage in such criminalized behaviors. Finkelhor and Asdigian posit that environmental science errs because it construes a particular range of criminalized behaviors as an appropriate basis for developing general, "universal" principles of victimization avoidance. Routine activities theory, for instance, is obsessed with stereotypical forms of offending, such as stranger assault and street crime.[31]

While acknowledging that victim blaming follows from a focus on stereotypical forms of offending and "overgeneralization," feminist scholarship elucidates how victimization avoidance is underpinned by a profoundly masculinist logic. Writing in the British context, Elizabeth Stanko explores police advice given to women concerning how best to avoid men's violence. As she notes, police advice typically suggests that women "should, and can, be individually accountable for our safety. To be a responsible woman is to display a healthy suspicion of men who appear to be ordinary men. . . . Such wariness is never-ending and resonates in questions we ask ourselves when we have been attacked: if only I did not walk home, open the door, shop at this store, and so forth."[32] According to Stanko, it is what is being left out of police advice material that is of greater significance; it "fails to condemn male violence as indicative of women's subordinate position in society. As such, it individualizes responsibility, without collective comment on *the problem of men*."[33]

On a related note, Sandra Walklate argues that conceptualizations of risk and danger cannot be divorced from the cultural ether in which they are developed. Furthermore, dominant constructions of risk transpire within an androcentric, masculinist culture. Concerning risks of criminal victimization, for example, the prevailing assumption is that exposure to such dangers can be controlled and managed via technology.[34] For Walklate, this way of framing the problem—seeking to control the unknown and turning to technical strategies to do so—evinces a masculinist bias.

Arguably of more relevance to the critique of environmental crime science, Walklate emphasizes that androcentric culture gives men license to seek out risky situations, while placing prohibitions on any desire that women may have for entertaining risk. This is evident insofar as women are perceived as having invited trouble—or are "deemed as 'asking for it'"—if they do seek out risky situations and something goes awry.[35] From this perspective, the masculinist logic that underpins environmental crime science surfaces when it questions whether victims remained alert, why they were out late at night, and so on. Such moments cannot but entail assumptions concerning who is authorized to take risks and who should be responsibilized for managing risk.

In our view, feminist critiques of victimization avoidance are not reducible to a dispute over whether crime control is best accomplished via a focus on offenders and punishment or on victims and crime-avoidance strategies. To be sure, we can understand why some might favor such a reading; feminist critique certainly shifts the gaze onto men's collective behavior rather than that of women. But this would be a fairly superficial reading, one restricted by a logic that regards crime as self-evident and thus limits dispute to how crime is best controlled. Central to feminist concerns are the normative assumptions embedded within victimization-avoidance advice and how these become an important aspect of narratives on crime and victimization. Victimization-avoidance advice and its corollary of victim blaming manifest the politics of environmental crime science: they operate as prescriptive forces that inscribe deficiency on some bodies (often those that occupy marginalized locations) while naturalizing the actions of others (often those from powerful social locations).[36]

Thus far, we have highlighted parallels between categories that provide neoliberalism with rhetorical defenses and recur, albeit in repackaged form, within environmental crime science. Both discourses posit that the state is incompetent, especially when compared to the efficiency that supposedly governs the private sector. Claims of efficiency entail the view that private interests can, and should, colonize services historically entrusted to the state. Lest this raise concerns, neoliberalism and environmental crime science insist that private actors operate with the greater social good in mind, which amounts to promising that whatever practices they instantiate will have beneficial flow-on effects. Finally, both discourses responsibilize individuals; they are adamant that the individual ought to bear the burdens associated with managing risk.

Neoliberalism, however, cannot be reduced to its rhetorical defenses. It also embraces and preserves social relations that are grounded in racism and classism. As do neoliberalism's rhetorical categories, environmental crime science replicates these aspects of the neoliberal order.

RACIST AND CLASSIST UNDERPINNINGS OF ENVIRONMENTAL CRIME SCIENCE

Much of the work that transpires within the orbit of environmental science is complicit with the racism and classism that structures neoliberalism. Moreover, this implicit racism and classism is routinely negated or downplayed in much of the critical scholarship. Indeed, one could read much of the literature in this area and quite reasonably conclude that there is an unspoken prohibition on suggesting that environmental crime science is complicit with racialized and class-based forms of inequality.

Published in 1982, James Wilson and George Kelling's short piece "Broken Windows" could be used to illustrate many of the deficiencies associated with environment-based approaches to crime control. It is, however, especially apt for demonstrating the racism and classism that pervade the field. Wilson and Kelling argue that the very functions of policing need to be overhauled. In attempting to make this case, they present an impressionistic account of the history of policing in the United States. In the first historical period, which apparently dates to "the earliest days of the nation," the primary function of police officers was to act as "night watchmen: to maintain order against the chief threats to order—fire, wild animals, and disreputable behavior."[37]

One may well wonder how police officers developed a conception of "order" during these "earliest days of the nation." Wilson and Kelling suggest that each "neighborhood" or "community" provided an appropriate image of order: "For centuries, the role of the police as watchmen was judged primarily not in terms of its compliance with appropriate procedures but rather in terms of its attaining a desired

FIGURE 10. Broken windows, abandoned buildings, and graffiti: signs of disorder and harbingers of serious crime, according to environmental crime science. It is perhaps worth noting that the aesthetics associated with BW theory echo those found within Lombroso's gallery of delinquents (cf. figure 2) and brain scan imagery (cf. figure 4). *Source:* Alamy, "Abandoned Apartment Buildings in Chelsea, NYC, USA" (S2RTAF), n.d.

objective. The objective was order, an inherently ambiguous term but a condition that people in a given community recognized when they saw it. The means were the same as those the community itself would employ."[38]

As in many other moments within the "Broken Windows" piece, Wilson and Kelling seem oblivious to the contradictions that plague their assertions. The meaning of order is "inherently ambiguous," yet "we all know it when we see it." In any case, this symbiotic relationship between the police and the "community" began to erode during the 1930s when "crime fighting," which relied on patrol cars and rapid responses to service calls, emerged as the core task of police forces.

Wilson and Kelling lament the rise of "crime fighting" as the primary goal of policing, claiming that it has ultimately failed to reduce crime while allowing neighborhood disorder to triumph. This leads them to advocate for a return to the "good old days" of foot patrols and entrusting police officers to rely on discretionary uses of their authority to enforce "community norms." This image of the past is disturbing not only for the faith that it reposes in police discretion, but also for its mythical construction of "community" and community-police relations. For Wilson and Kelling, neighborhood communities throughout the nation's early history were apparently bastions of consensus that inevitably forged good working relationships with police officers and were entirely devoid of any race or class tensions.

As numerous scholars have shown, however, this is hardly a plausible picture of the past.[39] Rather than being harmonious, relations between police officers and community members were often tense, if not overtly conflictual, in the pre-1930s era. Some posit that the police were often a major source of urban conflict precisely because they were not invested in serving the general public.[40] More important in this context is that to portray communities as homogenous and neatly bounded by shared collective norms is without foundation. If anything, neighborhoods lacked stability and were heterogenous in their composition. The exercise of police discretion and force routinely operated in ways that were consistent with social divides grounded in class and ethnic differences.[41] According to Frank Donner:

> A strong case has been made for the thesis that in the course of the past hundred years urban police have served as the protective arm of the economic and political interests of the capitalist system. What is especially compelling is that specialized police units have openly performed such functions. . . . These cadres have, beginning in the Gilded Age, predominantly engaged in political repression, which, in the context of policing, may be defined as police behavior motivated in whole or in part by hostility to protest, dissent, and related activities perceived as a threat to the status quo.[42]

To make a return to "community" modes of policing seem desirable, Wilson and Kelling must mythologize the past. In doing so, the same fate awaits the present. The convenient image of the past is used to suggest that "communities," supposedly bounded by some kind of normative consensus that transcends race and class contradictions, readily exist in the present and can be entrusted to regulate informalized, discretionary uses of police authority. The racism and classism of environmental science inhere in denying that race and class are forces that structure contemporary communities and neighborhoods. With the complexities of race and class relations denied, "community policing" can suddenly appear neutral or objective in its application.

Based on an ethnographic analysis of community-police advisory board (CPAB) meetings in Los Angeles, Aaron Roussell reveals the fractured nature of community and how it leads to policing that is far from neutral. Roussell finds that particular communities develop a sense of their boundaries by constructing "anti-communities," or outsider groups that figure as the source of problems. At CPAB meetings, Black Americans construct Latinos as the cause of community disorder; their disruptiveness is said to lie in the proliferation of unlicensed food carts and loitering (as they await day labor opportunities). Latinos, however, construct Black residents as the threat to community order. In their accounts, Black people are constructed as responsible for violence, and this is largely attributed to involvement in the drug trade and gangs.

For their part, the police tend to side with the Latino community, or at least that portion that participates in community policing meetings. Roussell attributes this to the broader political-economic forces that currently structure Los Angeles: Latinos may engage in illegal food vending, but they are perceived as hardworking, and their informal labor is pivotal to the exploitative dynamics of capitalism. By way of contrast, and not without some irony, Black Americans are shut out from contemporary labor markets because, at least in part, their citizenship status means they are not so readily exploitable. Dislocated from the economy and associated with drugs, gangs, and violence, they are successfully constructed by particular sections of the community (Latinos and police) as an "anti-community" and thus are subject to greater scrutiny, policing, and formal control mechanisms.[43]

The denial of community tensions and competing interests is not the only way in which BW evinces an underlying racism and classism. Wilson and Kelling underscore the supposed racial and class characteristics of the offenders who appear in the illustrations that buttress their argument. These intrusions into the text appear quite surreal at first glance:

> The car in the Bronx was attacked by "vandals" within ten minutes of its "abandonment." The first to arrive were a family—father, mother, and young son—who removed the radiator and battery. . . . Then random destruction began—windows were smashed, parts torn off, upholstery ripped. Children began to use the car as a playground. Most of the adult "vandals" were well-dressed, apparently clean-cut whites. The car in Palo Alto sat untouched for more than a week. Then Zimbardo smashed part of it with a sledgehammer. . . . Within a few hours, the car had been turned upside down and utterly destroyed. Again, the "vandals" appeared to be primarily respectable whites.[44]

What is fascinating about this passage is how white offenders appear as *counterintuitive* figures. We are supposed to be *surprised* by their presence. However, we can only be surprised by something insofar as it is unexpected. The revelations of whiteness suggest that Wilson and Kelling are aware that the BW argument will be read as racist and classist unless they somehow disqualify any such interpretation. And so, white offenders appear as a "plot twist." But this need to specify offender characteristics arises only because so much of the BW theory is racially coded and because of the broader context in which it is being written. Pointing out whiteness amounts to a preemptive strategy to deflect accusations of racism and classism that can quite easily be supported.

As a final example, Wilson and Kelling are compelled to admit that the vision of community policing for which they are advocating is not "easily reconciled with any conception of due process or fair treatment."[45] They portray the tension between informal policing and ensuring equitable treatment as a "serious concern," but are evidently more invested in consigning the problem to the "too hard basket": "How do we ensure that age or skin color or national origin or harmless mannerisms will not also become the basis for distinguishing the undesirable from the desirable? How to ensure, in short, that the police do not become the agents of neighborhood bigotry? We can offer no wholly satisfactory answer to this important question. We are not confident that there is a satisfactory answer except to hope that . . . the police will be inculcated with a clear sense of their discretionary authority."[46]

"No wholly satisfactory answer": we suppose this is what logically follows *after* one has argued in favor of policing practices that are premised upon a *disregard* for the constitutional protections and procedural rules that afford the public some degree of freedom from undue state intrusion, even if such protections and rules are nominal at best. What about some form of police accountability or oversight, perhaps even sanctions for racially biased policing practices if discerned? How about a little less hope and more skepticism concerning the idea that police will use their discretion wisely? Not surprisingly, these issues raise no blips on the Wilson and Kelling radar. It is incredibly disingenuous to promote a model of policing in which officer discretion is accorded a sacred place, recognize that this will open a door to discriminatory practices, but proceed to hope that relatively unrestrained police powers will not lead to abuses.[47]

We have devoted a considerable amount of space to the initial articulation of BW, but by no means is the racism and classism of environmental crime science limited to this one piece.

In much of the work occurring within the fold of RAT, one will find a series of "variables" and "measures" that pertain to the social composition of neighborhoods, such as "ethnic heterogeneity," "minority composition," "socioeconomic status," and "household annual income."[48] These are inevitably positioned as independent variables and thus are construed as possessing some value for explaining patterns in the crime types with which proponents of RAT are overwhelmingly concerned: interpersonal violent offending or property offenses.

The polarities or ranges that accompany these variables are what one might expect to find: a neighborhood will fall somewhere along a spectrum marked by "ethnic homogeneity" at one end and "ethnic heterogeneity" at the other; the residents of a neighborhood will be more or less homogenous/heterogenous in terms of their socioeconomic status. Once these variables are factored into hypotheses and measured, it is far from surprising to find that they account for some portion of the variance across whatever type of crime is being analyzed.

It is worth noting that such epistemic tendencies evince "white logic," a notion outlined by Kathryn Henne and Rita Shah and Nicole Gonzalez Van Cleve and Lauren Mayes that facilitated our analysis of actuarial science in chapter 3. As mentioned, white logic reduces race and class to static factors and attributes to them explanatory power. In doing so, there is a return to biological determinism, and the way race and class are actively constructed by criminological accounts and criminal justice processes is denied.[49] In environmental crime science, the variables may pertain to groups or neighborhoods rather than individuals, but the problem is much the same: the scholarly account is better understood as offering a construction of the variables that it portrays as "explanatory." That is to say, something like "ethnic heterogeneity" does not explain "higher crime rates"; rather, "higher crime rates" construct "ethnic heterogeneity" as problematic.

Consistent with its dependence on white logic, environmental crime science is driven to a defense of segregation. This is most evident in those studies in which it is claimed that neighborhood *homogeneity* renders an area less susceptible to crime. Of course, if homogeneity were refracted through a critical lens, one can almost imagine a discussion about social equality opening up: if we were to reduce class inequality and deflate the cultural authority granted "race" to structure social relations, then society would be "homogenous," and "crime rates" might go down. This, however, is not even remotely close to what RAT scholars appear to have in mind when they accord homogeneity an explanatory role in their theorizing of crime.

How, then, does the notion of homogeneity appear in these kinds of studies? In their analysis of burglary rates in The Hague, the

Netherlands, Wim Bernasco and Floor Luykx offer six hypotheses. In the present context, it is the third that is of most interest: "Higher levels of ethnic heterogeneity increase residential burglary rates."[50]

Ethnic heterogeneity appears among the hypotheses because the authors construe it as a force that reduces "territoriality." In the words of Bernasco and Luykx, "high levels of ethnic heterogeneity" undermine the ability of "neighborhood residents to get to know each other and integrate."[51] A neighborhood that lacks social cohesion and integration—"territoriality"—is supposedly more open to criminal activity. It probably goes without saying, but "ethnicity" is effectively treated as an ahistorical fact when framed in this manner. Moreover, it appears as a force that prescribes limits to what kinds of social relations are possible. "Cohesive diversity" is rendered oxymoronic, improbable, unimaginable.

After what would appear to be an excessive amount of attention devoted to outlining the study's methodology and data, as is somewhat customary of research in this style, the authors finally get around to stating that their most robust model reveals "strong support for all six hypotheses."[52] In concluding, they reiterate the importance of neighborhood characteristics in explaining burglary rates: "In line with the outcomes of previous research in the ecological tradition, lack of territoriality [i.e., high levels of ethnic heterogeneity and high residential mobility] and affluence appear to have rather strong positive effects on a neighborhood's burglary rate."[53]

Somewhat akin to Bernasco and Luykx, Andromachi Tseloni and colleagues list the "key implications" that follow from their comparative study of burglary victimization, closing with the following: "As a final point of domestic (UK) interest, there is a strong tendency to conflate area poverty with the need for crime reduction. This study reinforces earlier work (Trickett et al. 1995) that more affluent people in poorer areas are in particular need of crime reductive effort."[54] A few pages earlier, we are told why this is a "point of interest" for UK audiences but less relevant in the US context. Apparently, high-income US households are "less vulnerable [to burglary] due to lack of proximity to and accessibility by potential offenders."[55]

In plain English, the problem for affluent people in the United Kingdom is their proximity to the lower classes. As such, it is they who are in "particular need" of assistance. Somewhat mysteriously, the idea of "a group in need of assistance" does not appear to extend to those relegated to the lower rungs of the social ladder! Conversely, affluent people in the United States are afforded distance from offenders and thus are protected from burglary. The environmental crime scientist refers to "heterogeneity" or "homogeneity" but is effectively talking about segregation. In fact, Tseloni and colleagues admit as much when they state that "limited public transportation in the US allows for greater segregation by income than in Europe."[56]

And yet, rather than problematize segregation—or the spatialization of inequalities grounded in racism and classism—those working within environmental science employ it as a solution to crime problems. One could easily be excused for reaching the conclusion that such scholars are lamenting the absence of a strict ordering of neatly bounded neighborhoods premised upon class and ethnic homogeneity.

CRITIQUES OF ENVIRONMENTAL CRIME SCIENCE AND ITS "RACE PROBLEM"

The problematic of race, class, and environmental crime science has not gone unnoticed in academic literature. This is most evident in analyses of policing practices, such as OMP and the adoption of COMPSTAT, that emerged in the wake of the BW thesis. The attention given to BW, and the policing strategies associated with it, is not surprising. After all, of the particular theories that see a viable solution to crime in the reconfiguration of space, it was BW that entered the public imagination with some force and went on to become incredibly influential among political elites and police officials.

The critiques of newly emerging policing practices are certainly important, and they contribute much to our understanding of the deficits associated with environmental crime science. However, there are at least two disturbing trends in the critical literature. First, there is a tendency to suggest that *practical implementations* of BW betray the theory. We disagree that the logic of BW has been betrayed by its implementation. If anything, the opposite is more accurate: discriminatory policing practices are entirely consistent with the logic advanced by BW.

Second, scholars often suggest that the policing strategies inspired by environmental crime science run the risk of causing a legitimacy crisis for the state. Quite rightly, several critics have claimed that if new policing practices are too draconian, the public may come to see the state as a threat to justice, not its guarantor. Insofar as maintaining social order requires the cooperation and consent of the public, the preservation of order will be made all the more difficult if the state forfeits its legitimacy. There is something to this argument. Indeed, as draconian policing strategies intensify, the state *should* lose legitimacy. What this critique misses, however, is that the rise of draconian policing tactics to regulate bodies and space has not spawned a general legitimacy crisis for the state. This lack of crisis can be understood when we consider how BW policing strategies are differentially implemented across social space, or how they are situated in their broader sociocultural contexts.

Recuperating Broken Windows Theory

The work of Jeffrey Fagan and Garth Davies can illustrate our first concern.[57] New York City implemented OMP in 1994. Connected to OMP was the expanded use of "stop and frisk" searches of individuals. Patrol officers were encouraged to intervene if they suspected an individual of committing minor or serious crimes. Relying on a range of available evidence pertaining to "stop and frisk," Fagan and Davies examine whether the strategy amounts to "a form of policing that disproportionately targets racial minorities."[58]

Their findings consistently highlight that policing practices are shaped by race and class dynamics. For example, they find that in police precincts with small minority populations, stops of individuals from such groups are disproportionate. Conversely, whites are underrepresented in stop and frisk data even in places where they constitute the majority population.[59] Furthermore, one Black American citizen is arrested per 9.5 stops. For whites, there is one arrest per 7.9 stops. This suggests that Black Americans are more likely to be stopped for unfounded reasons.[60] Arguably of more importance in the present context, Fagan and Davies find that the racially skewed nature of "stop and frisk" searches cannot be explained by participation in crime. As they put it, "police over-stopped black and Hispanic citizens relative to their crime participation, well in excess of their white neighbors, and more often without constitutional justification."[61]

What their data and analyses ultimately show is that patterns in stop and frisk policing are very much determined by the level of poverty and social disadvantage that characterizes a neighborhood. Rather than being randomly distributed or governed by offending rates, the resources devoted to stop and frisk are concentrated and then directed to controlling the city's most economically marginalized members.[62] The racial dynamics to stop and frisk follow, given that class and race intersect in profound ways.

These findings paint a compelling picture of how stop and frisk policing amounts to a discriminatory practice. However, Fagan and Davies go to great lengths to argue that such racially skewed, class-based policing strategies are antithetical to the BW theory. They suggest that BW is at odds with stop and frisk policing for three main reasons. First, in implementing OMP, the New York Police Department (NYPD) concentrated on social disorder (i.e., people) rather than physical disorder. Second, whereas BW emphasized informal policing strategies, the NYPD refused to move away from "traditional" policing tactics that relied upon arrests and prosecution. Third, instead of working with "communities" to identify perceived problems within neighborhoods and devise novel strategies, the police adopted COMPSTAT to set policing priorities and institutional goals.[63] When taken together, it is these "deviations" from the "complex and nuanced theoretical origins" of BW that have led to the "policing of poor people in poor places."[64]

To be sure, there is some basis for portions of this reading of the BW theory. However, as our previous remarks intimate, it is also relatively easy to square the theory with criminalization of society's most marginalized members. We must admit, we find it bizarre that Fagan and Davies claim that the NYPD's decision to concentrate on social disorder somehow breaks with BW. By no means did Wilson and Kelling suggest that police should concentrate their attention solely on *physical* disorder. Their work is scattered with references to people and particular behaviors that they deem to be problematic. Wilson and Kelling claim, for instance, that many citizens are especially frightened by encountering disorderly people, such as "panhandlers, drunks, addicts, rowdy teenagers, prostitutes, loiterers, the mentally disturbed."[65] They go on to assert that for many people the experience of being confronted by an "obstreperous teenager or a drunken panhandler" is "indistinguishable" from "meeting an actual robber."[66]

More importantly, perhaps, proponents of BW often suggest that a distinction can be drawn between people and behavior.[67] This distinction is disingenuous; we cannot think of a single "behavior" that transpires in the real world independently of a person. No matter how blatantly artificial the distinction may be, it is necessary if one is seeking a mechanism that makes it possible to control certain segments of the population while neutralizing any criticism of such control. If we accept that people and behavior are distinguishable, it is inevitably the *behavior* that is being policed, never the *person*. More so, we can suddenly punish behavior without criminalizing an individual. Large swaths of the homeless population may be asked to "move on," or be "arrested," or be issued a ticket for loitering, but this suddenly has no connection to their class status. If they want to avoid police harassment, all they need to do is stop engaging in "homeless behavior." And of course, most of the behaviors that BW proponents see as "disorderly" have obvious class dimensions: panhandling, loitering, drug/alcohol abuse, and so on. Given these moments, how BW can come to be regarded as not providing a warrant to inscribe criminality on "poor people in poor places" is perplexing.

Fagan and Davies are on slightly stronger ground when they suggest that BW emphasizes informal policing.[68] This, however, is counterbalanced by Wilson and Kelling's insistence that it is unwise to decriminalize, or neglect, disorderly behaviors. As they put it: "The wish to 'decriminalize' disreputable behavior that 'harms no one' . . . is, we think, a mistake. Arresting a single drunk or a single vagrant who has harmed no identifiable person seems unjust, and in a sense it is. But failing to do anything about a score of drunks or a hundred vagrants may destroy an entire community. A particular rule that seems to make sense in the individual case makes no sense when it is made a universal rule and applied to all cases."[69] If anything, this logic of contagion ordains that "traditional" police tactics—arrest and prosecution—may be deployed not so much for the sake of controlling any given individual "offender," but for the control of groups of people.[70] Wilson and Kelling even admit that charges like vagrancy and public drunkenness have "scarcely any legal meaning."[71] Apparently, however, they exist because "society" wants police officers to have a legal mechanism to "remove undesirable persons from a neighborhood when informal efforts to preserve order" have failed.[72] As such, the strategies of arrest and prosecution are not incongruent with BW. Rather, they are the real possibility of *force* upon which the effectiveness of "informal" policing strategies hinges.

Finally, there is the argument that COMPSTAT emerged to inform policing strategies, thereby foreclosing input from the community about appropriate "neighborhood standards" or tolerance thresholds. Again, there is perhaps some basis to this. It is nevertheless rather problematic in our view. As intimated earlier, Wilson and Kelling assume that insofar as officers derive their sense of order from the "community," their policing tactics will remain fair and impartial (and this even if constitutional protections are violated). Of course, what they expunge from the historical record, and what they refuse to acknowledge about contemporary social relations, is that racism and classism are powerful structuring forces within and across communities.

This tendency to gloss over sources of social conflict leads to a mythical notion of community, one in which a normative consensus surrounding acceptable behavior, and responses to it, triumphs. We doubt that such communities can routinely be discovered, if at all. At a bare minimum, it would have to be conceded that those who belong to the social groups most susceptible to BW policing are unlikely to recognize it as somehow serving their best interests or in line with their standards. In a society fractured along the lines of class, race, gender, age, and so on, there is no such thing as "community." There is only this or that faction of a community.[73]

COMPSTAT may have originated from within police departments, but it has hardly been met with mass resistance from the public. In fact, some scholars have demonstrated that COMPSTAT, relative to other technological innovations, can easily be regarded as having experienced an unusually rapid rate of diffusion.[74] As such, we are not convinced that COMPSTAT somehow manages to "take the policing world by storm" while remaining impervious to "community" expectations. If anything, COMPSTAT appeases that portion of the public with a pronounced desire to be informed of police activity, who especially wish to see intensified modes of policing combined with drops in crime. These are precisely the kinds of figures—arrest rates and crime rates—that COMPSTAT is designed to spit out.

Our point in all of this is not to get bogged down by excessive detail, but to raise a question: Why go to such lengths to recuperate a theory like BW? Why not simply call the theory out for its racist and classist underpinnings, its complicity with draconian policing tactics? We take it as an indication that the ideological tenor of environmental crime science, particularly its denial of racism and classism, has been internalized and granted consent within much of mainstream (administrative) criminology. That is to say, the kind of conservatism espoused by the Wilsons and Kellings of criminology is hegemonic, so much so that to transgress or interrogate that thought is to violate an unspoken cultural taboo of sorts. Problematic practices may be interrogated, but the broader discursive regimes that provide the warrant for those practices are "off limits."

The Legitimacy Crisis That Never Happens

Our second concern revolves around the theme of "legitimacy" and how it appears in critiques of place-based policing strategies. Several scholars suggest that the intensified use of policing tactics grounded in environmental science (i.e., "order maintenance," "zero tolerance") may undermine the state's legitimacy.[75] This argument hinges upon demonstrating that new modes of policing are excessive, if not draconian. The logic appears to be that policing tactics that amount to excessive intrusions upon the private lives of citizens—perhaps one should simply say "authoritarian" policing strategies—will not be much appreciated, thereby leading to corrosion of public trust in the state.

The draconian part of this equation is relatively easy to demonstrate. Drawing from Freda Solomon, K. Babe Howell compares figures pertaining to the volume of arrests in 1989 and 1998 in New York City.[76] These figures show dramatic increases in the number of arrests among individuals with no prior criminal record, and that the proportion of white people arrested dropped, whereas the proportion of Hispanic people arrested increased.[77] Most of this arrest activity centers on misdemeanors, especially marijuana possession, increasing from 9,208 arrests in 1989 to 43,540 in 1998.[78] Consistent with Fagan and Davies, Howell also notes that this policing is racially skewed: of misdemeanor arrests in New York City between 2000 and 2005, 86 percent were of Black (48–50%) and Latino (approximately 27%) individuals.[79] Misdemeanor arrests may sound trivial to some, but because they generate criminal records and often very brief prison sentences, they can easily sabotage the life chances of individuals.[80]

Others have documented how OMP often means that constitutional rights and protections are ignored by officers in their interactions with citizens.[81] If citizen complaints against police are anything to go by, this claim is well founded. Not long after the introduction of BW style policing, civilian complaints concerning the excessive use of force by police shot up by 61.9 percent; complaints in relation to illegal searches jumped by 135 percent; and claims that police had illegally searched an apartment rose by 179 percent.[82]

The loss of legitimacy, however, is not so easy to support. Tom Tyler's work on procedural justice demonstrates that, contrary to the instrumentalist assumptions of the deterrence model, people obey the law not because they are afraid of being caught and sanctioned, but because they believe the law is fair and the authority of those imposing the law is legitimate. If those who enforce the laws are viewed as illegitimate, they can lose the support of the public.[83] With this in mind, scholars suggest ways to prevent the loss of legitimacy. Many of the proposed ways to reform the criminal justice system, such as decriminalizing a range of minor offenses or expunging criminal records after a specific time period, certainly make sense.[84] Other proposals to protect the state's legitimacy are not necessarily problematic, but they do appear to be premised upon an undue optimism. John Eterno and Eli Silverman suggest that individuals need to take seriously the sworn oath that must be made to become an officer, which entails promising to "support and defend the Constitution of the United States" in the course of performing their duties.[85] They also emphasize the need to recognize that *reducing crime while protecting Constitutional rights* is more important than simply *reducing crime*.[86]

Making these kinds of suggestions is all well and good, but very few scholars seem invested in drawing attention to what would appear obvious: despite all the absurdities and injustices that accompany policing practices inspired by environmental crime science, there is no *general* legitimacy crisis. Moreover, it is this *lack of crisis* with which we should be concerned and that requires explanation.

Although this problem has eluded many of those invested in interrogating new policing strategies, its resolution is embedded in the studies they carry out. What the studies show is that the draconian controlling of space is not implemented in a random manner, but is shaped by class and race. The same kinds of figures can be released year in and year out: stop and frisk data that show the perpetual targeting of marginalized individuals in marginalized communities, the disproportionate arrest of marginalized individuals for trivial behaviors that are nevertheless criminalized, such arrests followed by excessive stigmatization, and excessive force complaints that overwhelmingly arise from incidents in marginal communities or from marginalized individuals. And yet the legitimacy crisis remains only a possibility; it is only on the far horizon that we catch a faint glimpse of its anticipated eventuation.

That the state's legitimacy is not in peril would be much more obvious if scholars worked with a pessimistic understanding of contemporary sociocultural structures. Such pessimism would generate a very different set of conclusions: environmental crime science and the policing practices with which it is associated are popular and receive support among the majority population because they are congruent with current political arrangements.

One need only present a counterfactual to illustrate the investment that dominant social groups continue to make in draconian modes of policing and to show that environmental crime science is socially recognized as a body of ideas that effectively sanctions the targeting and control of specific populations: if 86 percent of the misdemeanor arrests in

New York occurred in neighborhoods like Riverdale, or in the vicinity of Grammercy Park, and if middle- and upper-class whites were overrepresented in this figure, one can only imagine the "public" opprobrium that would be directed against the police force and elected officials. Likewise, one can only imagine the sudden number of lawsuits that would be taken more seriously, the number of disputed hearings that would clog the courts, and so on. Some may counter this by noting that there has been a spike in complaints against the police. Yes, and yet the same patterns recur. What seems to matter, therefore, is *who* is complaining.

CONSTRUCTING "OFFENDERS": THE BIO-ECOLOGICAL IMAGES WITHIN "RATIONAL-CHOICE" AND THEIR PUNITIVENESS

Environmental crime science often posits that the "motivated offender," alongside a "victim" and the "absence of guardianship," constitutes a necessary ingredient for any given crime event to transpire. Somewhat strangely, the concept is often excluded from consideration at the very moment of its introduction. In Cohen and Felson's "Social Change and Crime Rate Trends," we need not read past the second page to find the motivated offender being disavowed: "[We] do not examine why individuals or groups are inclined criminally, but rather we take criminal inclination as given and examine the manner in which the spatio-temporal organization of social activities helps people to translate their criminal inclinations into action."[87] It would appear that this exclusion has since emerged as a leitmotif of environmental crime science. In their examination of burglary, Bernasco and Luykx note: "The main assumption underlying our model is that it takes the existence of motivated burglars for granted. It does not attempt to explain how people become burglars, it attempts to explain why they burgle where they burgle, instead of somewhere else."[88]

That the disavowal of motivation is common within environmental crime science is perhaps best illustrated by Clarke, who considers whether Israel's construction of the West Bank barrier amounts to an effective "target-hardening" strategy against Palestinian attacks. Clarke notes that a wide variety of international organizations—including the United Nations, the International Court of Justice, and Amnesty International—have "declared the barrier illegal."[89] It has been declared illegal by such organizations insofar as it establishes borders between Israel and Palestine, effectively "annexing large parts of the West Bank and all of East Jerusalem."[90] Despite the illegality of the wall, let alone its broader geopolitical context, Clarke's emphasis on target hardening leads him to willingly ignore "political controversies surrounding the barrier" and assert that we need not "waste time on motives."[91]

It is odd to stumble upon a theory that posits some *necessary condition* to that which it purports to explain, but then proceeds to disavow any in-depth exploration of it. In our view, the tendency to bracket out questions concerning motivation occurs for at least three reasons. First, it is consistent with the atheoretical nature of environmental crime science. Second, and closely related, it follows from the presumption that such questions have been sufficiently addressed via "rational choice" and can therefore be closed. In this sense, it is as if environmental science thinks that theory is now redundant and so we can proceed to implementing technical fixes. Third, rhetorically jettisoning motivation allows environmental crime science to surreptitiously offer constructions of offenders as it devolves into a litany of quasi-experimental reports concerning how this or that modification of space impacted some particular criminalized behavior. These constructions justify the adoption of punitive logics and practices.

In these report-like writings, it becomes apparent that the purported grounding in rational choice is quite misleading: the actor is conceptualized in ways that drastically differ from such a construct. At one end of the spectrum, some offenders are construed as weakly committed to criminalized behavior but willing to take advantage of opportunities nonetheless; at the other end are those who are strongly committed, so much so that they may put much energy into outwitting preventive environmental measures. That rational choice can be said to encompass such disparate "motivational structures" is to invest it with an elasticity that is difficult to justify. The concept becomes rather meaningless because it construes any manifest behavior as evidence of choice.

Rather than simply accept its proclamations, environmental crime science is better understood as wedded to "bio-ecological" conceptions of behavior. In this, it proffers a typology consisting of two offenders: those who are "opportunistic" and those who are "habitual."

"Opportunism" is the conception of offending that appears most often. In this view, the offender is much like a Darwinian organism, "selecting" behaviors that are adaptive or "fitted" to their immediate environment.[92] Insofar as opportunism implies weak motivational structures, every individual is thought capable of falling into this category. As Clarke writes, because "situational prevention sees everyone as susceptible to crime opportunities," it does not "draw hard distinctions between criminals and others."[93] The "habitual offender" is rarely articulated directly; as a result, the problem it poses for environmental crime science is not addressed. Nevertheless, it is constantly posited in studies and evaluations of preventive measures as the residual that never goes away. The habitual offender—or something very much like it—is suggested by the volume of crime that continues despite situational interventions.

That the habitual offender haunts environmental crime science reveals a profound contradiction within the approach. On the one hand, positing a weakly motivated, opportunistic offender implies that the elimination of crime is not only possible but a relatively straightforward, technological affair. One might go so far as to say that it amounts to a truism: the opportunity thesis asserts that if the environment is modified such that crime cannot be enacted, then crime cannot be enacted. This is no different than arguing that the consumption of sugar can be prevented by removing it from the aisles of supermarkets. Or, if we want to avoid gun-related fatalities, prohibit the production of guns. If we want to kill a species, eliminate its food source. On the other hand, however, this truism is constantly refuted as its proponents strive to demonstrate it: many SCP studies, for example, reveal some portion of offending that simply refuses to dissipate. In order to rescue itself from this seemingly unperturbed "residual," environmental science is compelled to reconfigure its primary goal as the "amelioration"—rather than "elimination"—of crime.[94]

To return to our main thread, deploying rational choice to submerge these constructs of offenders normalizes punitive interventions. This holds despite rhetoric to the contrary, such as Clarke's suggestion that SCP, mysteriously in his view, "lacks a natural constituency among politicians." Those of liberal persuasion, apparently, should be "attracted to its essentially non-punitive philosophy."[95] Disclaimers aside, the implausible elasticity attributed to rational choice responsibilizes all individuals by asserting that each particular behavior is animated by some degree of contemplation. The bio-ecological conceptions of offending then push this responsibilization even further, albeit in slightly different ways.

Insofar as environmental crime science posits a subject that is weakly committed but summoned to act by situational factors, it denies that there is any real or significant motivation for crime. This leaves commonsense understandings of "crime" undisturbed while rendering nonsensical legal notions and sociological narratives that create room to consider the contexts of offending behaviors. The legal concept of "mitigating factors," for example, typically allows an individual's social, economic, and cultural circumstances to be considered within criminal justice settings, but is made redundant. Much the same could be said for critical sociological concepts. The standard drivers of crime found within sociological narratives—"poverty," "inequality," "relative deprivation," "strain," and so on—are repressed by the "opportunistic offender."[96] In closing their often-cited piece on routine activities, Cohen and Felson suggest that the problem may well be that the contemporary "structure for legitimate activities" is too generous: "Rather than assuming that predatory crime is simply an indicator of social breakdown, one might take it as a byproduct of freedom and prosperity as they manifest in the routine activities of everyday life."[97]

Given that strongly committed offenders are the exception rather than the rule in environmental science—if not a "by-product" of "prosperity"—contemporary social relations are rendered "nonmotivational" and therefore "acceptable." That the majority of offenders are made to appear weakly committed and easily able to fit themselves into the social order if crime opportunities were blocked indicates that "mitigating factors" are not much more than "weak will" and "pathetic excuses." Moreover, because all individuals are thought to be opportunistic and thus embody some propensity for engaging in criminalized behaviors, the perspective can assert that its promotion of punitiveness is nondiscriminatory and claim that it subjects all equally to its logic. Indeed, if situational interventions suppress opportunities that any individual might exploit, then they cannot be discriminatory by default.

There is an irony here. One might just as easily read environmental crime science as revealing a form of "entrapment": situational factors are described as summoning criminalized behaviors, as if spatial vulnerabilities were akin to dangling a carrot in front of a horse. In this, behaviors are presumed to be amoral; they simply follow as "rational actors" encounter their environs. And yet the environmental crime scientist proceeds to assert that some courses of action should be rejected, that our opportunist should operate with *moral sensibility* (so much for the reduction to rational choice). Rather than responsibilize designers who would seem to lack foresight and rational judgment, the horse is blamed for taking the carrot. Given the insistence upon opportunism—the notion that everybody (and therefore nobody) is a "criminal"—how behaviors are demarcated and where blame is assigned appears arbitrary, an imposition that follows from political standpoints.[98]

The implicit positing of habitual offenders suggests that some are simply stubborn; despite situational controls and modifications of the environment, their commitment to behaviors deemed criminal persists. Paradoxically, constructing most offenders as opportunists is to insist that design modifications should rest content with reductions in crime, rather than eradication, and this in turn leads the environmental crime scientist to confine to the shadows the habitual offender that it cannot but intimate. In this, the type of offender that other crime sciences presuppose, reinscribe, and typically obsess over ceases to be of concern for environmental science; it is ignored as a matter of convenience. For instance, whereas actuarial science seeks to identify and preempt the movement of habitual offenders in many respects, environmental science posits their presence while demanding their discursive absence. Were this "habitual offender"—or the residual that remains despite situational interventions —allowed to step out from the shadows, environmental science would implode; the enterprise reveals something that it cannot account for despite construing itself as a general theory on crime and its control. And yet this residual is certainly pathologized, given that its resilience transpires against a presumption that it ought to be "ameliorated." At some point in the narrative, the aporia inserted in rational choice cannot but interrupt the text: "We may all be prepared to steal small items from our employers, but few of us would be willing to mug old ladies in the street."[99] Suddenly, not all individuals can be encompassed by a single category; some are rendered different from the rest of us.

In short, environmental crime science prescribes either that there is no "real" motivation to commit crime (the opportunist), or that crime is the province of subjects strongly motivated to outsmart situational measures (where such measures are invariably presumed sensible and thus beyond question), or—somewhat obliquely—of defects in moral reasoning (these last two prescriptions being embedded in the habitual offender). The former type is rendered inexplicable by presupposing that another path of noncriminal action could easily and always be taken; the latter is pathological in its "deviance" from the "opportunistic norm," apparent in the rare form of rationality and moral judgment that it is thought to exercise. Bridging the inconsistent meaning attributed to rational choice, neither actor can be comprehended—one is "stupid," the other "evil"—and thus both are liable to technological suppression and punitive intervention.

REDUCING OPPORTUNITIES FOR CRIMINAL JUSTICE AND CRIMINOLOGY

We have had occasion to note in passing that environmental crime science claims to advance a general theory of crime and its control. At their most abstract, theoretical level—a strange phraseology to use in this context—environmental approaches rely on categories such as "rational actors," "opportunities," and "targets" or "victims." In RAT, for example, constructs along these lines—offender, lack of guardianship, victim—are posited as basic ingredients of all crimes. It follows that if any one of these "necessary conditions" can be eliminated, then crime cannot transpire.

It is difficult, however, to reconcile claims of generalizability with the particular concerns or "empirical" subject matter by which environmental crime science is saturated. This gap between assurances to offer a general theory and substantive foci can be discerned in at least two ways. First, environmental crime science is noteworthy for the forms of criminalized behavior that rarely enter its field of vision. The second edition of Clarke's *Situational Crime Prevention* contains twenty-three case studies, which accompany the "theoretical" introduction. The principle that supposedly governs the inclusion (and/or exclusion) of studies is "the need to demonstrate the generality of situational prevention" by covering a range of "environmental contexts and offenses."[100] These "contexts and offenses" are then summarized: "Environmental contexts covered include private homes, shops, post offices, convenience stores, parking facilities, public telephones, street markets, night life and red light districts, leisure complexes and different forms of public transport. The offenses covered include auto thefts, welfare fraud, toll fraud, burglary, robbery, employee theft and shoplifting, drunkenness and assaults, vandalism, graffiti, soliciting, fare evasion and obscene phone calling."[101]

As this list intimates, white-collar crime, business or corporate crime, and state crime—in short, criminalized behaviors associated with the powerful—are entirely absent. The "employee theft" amounts to a case study of how tracking inventory led to reductions in the volume of stolen merchandise, such as VCRs and camcorders, within a large electronics retailer.[102] The exclusion of criminalized behavior more readily associated with the powerful is further reinforced by those moments in which the motivation for research is described. In evaluating the electronic tagging of merchandise to inhibit shoplifting, for example,

Robert DiLonardo notes that the work "was undertaken with the full cooperation of, and for the sole benefit of the retail chains which participated."[103]

A second way to frame the gap between promising to offer a general theory and substantive research practice entails identifying the particular construct of crime with which environmental approaches effectively operate. In this sense, it is not unreasonable to posit that environmental crime science routinely presupposes two elements in its working definition of "crime." On the one hand, crime is reduced to its stereotypical forms (i.e., property crimes and interpersonal violence, especially violence involving offenders and victims that are strangers to one another); on the other hand, for a criminalized behavior to find its way into environmental crime science, it often needs an element of concentration in time and/or space (e.g., a concentration of robberies on some particular block, as occurs in hot spot policing; theft from auto vehicles in this or that car park, as in SCP; recurrent graffiti in an area or on some surface, as in CPTED).

There are, however, other ways of constructing the category of "crime." One may just as well emphasize criminalized behaviors that are prevalent but dispersed (e.g., intimate partner or family violence; some forms of men's sexed crime against women, such as date rape or sexual assault of an acquaintance), or perhaps criminalized behaviors that are less prevalent but devastating in their consequences (such as financial crimes like Ponzi schemes or corporate crimes that devastate ecosystems—the Bhopal disaster of 1984 and the BP oil spill of 2010 come to mind).[104]

Of course, insofar as the "general theory" of environmental crime science is not much more than a truism, as we have intimated, it could be applied to financial and corporate crimes: to prevent massive oil spills that more or less devastate an entire ecosystem, simply ban deepwater oil drilling. Such statements, or extensions of environmental crime science's logic to the criminalized behavior of the powerful, are, however, hard to find. In saying this, we are not suggesting that the problem is one of empirical focus and that environmental crime science would be improved were it to extend its gaze. Rather, its omissions—alongside the things it chooses to include—reveal the political logic by which it is informed and which it reproduces as a discourse. To put it bluntly, environmental crime science embodies neoliberal subjectivity and operates as a handmaiden to neoliberal capitalism; it obsesses over controlling the powerless while striving to bolster the unrestrained freedoms of the powerful.

What is somewhat peculiar about environmental crime science is its unabashed attempts to marshal this political logic into dictating the parameters of crime control and criminology. To be sure, other crime sciences do participate in this kind of endeavor; biological crime science, as discussed in chapter 2, increasingly insists that criminology must factor "biology" into its theoretical accounts if it wishes to secure its status as a "science." Environmental approaches, however, appear to be far more explicit, encompassing, and prescriptive in their vision.

The previous sections of this chapter make clear that environmental crime science seeks to reduce crime control and criminal justice to technological interventions that suppress criminalized behaviors (particularly those that meet the criteria of being stereotypical and concentrated). In this, incapacitation is taken for granted and fetishized. Moreover, the idea that "criminal justice" should remain the province of the state is corroded as private actors and their interests are construed as delivering more efficient technologies and strategies of crime control. That such actors are unaccountable to the public appears to be of no concern to the environmental crime scientist.

Not surprisingly, the political logic of neoliberalism resurfaces in efforts to prescribe limits to criminology. It is worth quoting Clarke at some length:

> Criminologists will need to define their theoretical goals more in terms of control than enlightenment, and will need to define control more in terms of reducing opportunities than propensities. They will have to become familiar with a host of social institutions—schools, factories, hospitals, rail and bus systems, shopping malls and retail stores—beyond the courts and the prisons. They must no longer disdain the business world, but must recognize its central role in the production and control of crime. Their role models will increasingly need to become traffic engineers and health specialists—professionals employed to improve everyday life—rather than academics and social commentators. In short, a more down-to-earth, pragmatic approach will be required.[105]

It is readily admitted that the pragmatism of environmental crime science "conflicts with the philosophy and values of criminology students, many of whom aspire to be social reformers seeking a reduction in inequality and deprivation."[106] However, the criminology student is assured that "equally rewarding and challenging careers await those who can shift their professional goals from long-term social reform to making an immediate reduction in crime—which, after all, harms the very people they seek to help."[107]

Like Clarke, proponents of CPTED also insist that short-term "immediate solutions" should be accorded priority. Long-term solutions can be entertained, but such a task is assigned to others: "We concentrate on changing the environment to minimize the opportunity for the offender to cause damage, rather than trying to change the offender's character or motivation. The attraction of this approach is that it can work in the short term while research and policy makers work on longer-term solutions to the problem of crime."[108]

As imagined by environmental crime science, then, criminology is to be informed by particular normative orientations, institutional allegiances, and subject positions. Rather than seek "enlightenment," one should place greater value on control and suppression; the criminologist should align with other sites of discipline and docile bodies (schools, factories, commercial interests, and the business world); and criminology should produce subjects who see themselves as technical specialists focused on making practical contributions to the world as it is rather than critical academics or intellectuals.[109]

In advocating for a "pragmatic" criminology, however, it is necessary to situate it against an alternative, and environmental crime science selects a criminology focused on entrenched problems and long-term solutions. This choice is not accidental. One might, for example, just as well construe a criminology dedicated to restorative justice as an undesirable alternative. But it would seem as though a critical, sociological criminology of the Marxist variety figures as impractical and affirmatively harmful. This propels the environmental crime scientist to prioritize short-term fixes over long-term solutions. The problem is neutralized by claiming that technical measures to suppress crime will ensure that those in impoverished circumstances are spared criminal victimization. The moral conscience of environmental crime science is pacified by assuming that social deprivation and a life of poverty are tolerable provided they are not accompanied by occasional, if not rare or unlikely, criminal victimization.

Moreover, this approach cannot but entail the additional presumption that impoverished social classes can be divided into their noncriminal and criminal counterparts, deserving and undeserving, worthy and unworthy. Once again displaying its condescending and paternalistic tone, environmental crime science is held capable of swooping in to save the former from the latter. In this, there is no recognition that environmental crime science imposes its own harms, or that its interventions might be, at the very least, double-edged: it suppresses and blocks possible lines of flight while doing nothing to open up new, "long-term" possibilities. If anything, it places prohibitions on the idea of expanding opportunities that would ease the deprivation and inequalities that it acknowledges. That, after all, is posited as the wrong kind of criminology to pursue.

Like the other crime sciences that have been examined, environmental crime science ultimately ends on a similar note: its goal is to maintain power asymmetries while promising to control individuals and groups that the contemporary political order deems undesirable. This preservation of order is chalked up as working to the benefit of all, despite admissions that current social arrangements are plagued by "inequality and deprivation."

Conclusion

Over the course of this book, crime science has been treated as a discourse. This is not to suggest that practice can be ignored; quite the opposite. To accord discourse the warrant to govern analysis hinges on recognizing that it is intimately bound to practice. As chapter 1 put the point, discourses and practices are always co-constituted: discourse harbors practice, and practice presupposes discourse. Insofar as they are so thoroughly intertwined, analysis can focus on how statements instantiate or revitalize practice, inflect concepts with new meanings, normalize power asymmetries, and so on. However, such an approach also allows one to consider how practices embody, convey, reproduce, or recraft discursive frameworks.

We have applied this approach across some of the branches and divisions that constitute crime science, most notably biological, actuarial, security, and environmental logics. Although these variations require discrete analysis and interpretation, there are features that reverberate throughout each—and thus throughout crime science broadly conceived.

Each manifestation ontologizes crime, treating it is a self-evident category. The tendency is perhaps most evident in Lombroso's strange, almost throwaway remark that ancient peoples were right to punish animals for dangerous behavior, which presupposes that the distinction between criminal and noncriminal can be found in, and thus derives from, the natural world.[1] Once crime is ontologized, it is but a small step to embracing representational epistemology, wherein the empiricist crime scientist accepts that there is a "real world" that transpires independently of the capacity to observe and, moreover, that it can be understood, theorized, and mapped, its "laws" of "cause and effect" discovered and exploited. In claiming the ability to reflect reality without mediation and to parse what is cause from what is effect, crime science invariably posits that some quality of the individual constitutes an "independent variable." For its part, crime figures as a "dependent variable." Arranged in this manner, crime comes to be understood as inhibitable by interventions that ultimately have as their point of application the individual, or other precursors to crime that enter an assemblage involving the individual.[2]

By reducing the causality of crime to forces that are readily observable rather than abstract—individuals, criminal records, architectural design, physical disorder, and so on—the promise that it can be eradicated with particular technological strategies or the right mix of technological interventions seems to follow almost automatically. It comes as little surprise, then, to find crime science fetishizing technology. It always has, even in the early work of Beccaria and Lombroso. The presumption that technology offers the most effective means to address what are construed as crime problems, however, is constantly betrayed by the particular technologies advanced and developed within crime science. Broadly speaking, crime science's technological interventions typically presuppose the persistence of those behaviors that they deem criminal. We have seen this throughout this book: DNA technologies assume and necessitate the crimes they retroactively help to prosecute; risk assessment needs data points to identify correlations; security efforts might keep crime at bay, but we can never be sure that threats have been eliminated; and environmental interventions reveal a residual criminality, although their logic dictates its extirpation. Moreover, technological interventions often reframe everyday behaviors as socially deviant, if not criminal. In some cases, the implementation of crime science technologies involves the transgression of legal and/or moral boundaries.[3]

Crime science—in ontologizing crime and construing it as vulnerable to technological control—is intimately connected to, and ultimately reinforces, power asymmetries. As Lombroso put it, "Fortunately, my scientific findings, far from making war on social order, reinforce it."[4] This becomes especially evident in the performative effects of crime science. Its discursive and technoscientific manifestations promote particular conceptions of crime, criminality, just strategies of crime control, law and jurisprudence, and, inter alia, criminology. To put this somewhat metaphorically, crime science is replete with conceptualizations of good and evil, worthy and unworthy, and with images of what a just social world would look like. In these conceptualizations and images, crime science casts itself as the purveyor of the greater social good, a position it usurps by promising to control "evil." Importantly, never does it suggest transcending its binary of good and evil. Instead, it insists upon the existence of evil and the need for it to be censured, neutralized, and eradicated. In this, it has no interest in interrogating contemporary political relations and power asymmetries; rather, it is wedded to their preservation.

To be sure, we use power asymmetries in a fairly broad sense to encompass disparities grounded by race, class, gender, sexuality, ableism, religious orientation, and so on. Of course, crime science need not operate along any one particular dimension; it may just as easily reinforce disparities that hinge upon the intersectionality of power relations. Having said that, one of the asymmetries that has repeatedly surfaced throughout this book is that organized by race. The racism embedded within crime science can be traced at least as far back as Lombroso, who unashamedly posited that there was no meaningful distinction between the Black body and criminality, thereby rationalizing its permanent incapacitation.[5] The thread of racism then runs throughout the crime science enterprise. Bio-forensic technologies produce racially skewed databases, and their sibling, biosocial criminology, summons the spirit of Lombroso to once again conflate criminality and the Black body.[6] The predictor variables so central to actuarial science are not hard to decode as proxies for race and class and, like bio-technologies, actuarial science's databases could hardly be described as "immaculate."[7] The "state of exception" instantiated by security science unleashes its most disturbing horrors beyond the western, Anglophone world but also within its "dark sites," which can exist anywhere even if they are always nowhere. In environmental crime science, social marginality is recoded as "disorder" that portends "serious crime" and civic collapse, thereby normalizing anti-constitutional, draconian modes of control.[8]

Although it involves numerous generalities, crime science is not a monolithic entity. Rather, its particular factions can be underpinned by disparate assumptions and loaded with discrete performative effects. We do not want to recapitulate all of the interpretations that we have offered throughout this text, but it is perhaps worth highlighting some of the ruptures and nuances across crime sciences. The dimensions noted earlier—crime, criminality, control, law and jurisprudence, images of criminology—can be used to summarize some of the subtle variations across the branches of crime science.

In addition to ontologizing crime, crime science often frames particular behaviors as deviant or, in some cases, recasts everyday acts as criminal. Moreover, crime science proffers various cartographies or ways of mapping crime. Biological science, for instance, conceptualizes those who treat rights to privacy with reverence as socially deviant. This is especially evident in relation to DNA technologies, which promote the view that we should be more invested in providing proof of innocence than in tasking the state with establishing guilt.[9] Security science has criminalized jokes and sarcasm (everyday resistances) at airport security checkpoints.[10] It also reterritorializes crime types according to what its technologies can observe and measure, which rationalizes the devotion of resources to a strange mix of criminalized behavior. Although not unrelated, environmental crime science also brings a strange mix of behaviors into the spotlight, but it relies on distinct conceptual strategies to do so. It draws a line that links "minor" and "serious" crimes, thereby justifying a focus on the former. Its mapping of crime also tends to prioritize forms of offending that are concentrated in time and space.[11]

All of the crime sciences traced over the course of this book locate criminality in the body in some way. By no means, however, are the body and its supposed deficits portrayed in a uniform manner. Bio-forensics asserts that crime can be detected, attributed to its author, and punished with the help of biometric technologies. The capacity to identify criminals is thought to be communicative and thus capable of deterring other would-be offenders. In this, a self-interested, rational actor is presupposed. The academic strands of biological science—that is, biosocial criminology—continue to theorize criminality in a Lombrosian sense, or as a "maladaptive adaptation" to environmental conditions.[12]

Actuarialism is less invested in directly binding biology and criminality. Rather, it treats criminality in probabilistic terms, construing it as a function of criminal records, history of drug and alcohol abuse, marital status, employment status, and so on. In our view, this is ingenuine and elides the ways in which its risk assessment instruments can just as easily be said to work in the other direction: rather than predict criminality, its instruments impute criminality to its predictors, which more or less encode social marginalization.[13]

For the environmental crime scientist, offenders are dichotomized: on the one side, they are seen as weakly motivated, opportunistic actors; on the other, they figure as "habitual," stubbornly persistent, or strongly motivated. The weak motivation thesis is used to assert that environmental controls ensure that crime will not innovate or be displaced. It is also used to assert that environmental approaches serve the best interests of those it seeks to control; if anything, would-be criminals will be channeled along normative paths from which they will benefit. The habitual criminal is evident as the residual that does not go away despite spatial modifications intended to eliminate crime. However, because it cannot be reconciled with claims of weak motivation or accounted for from within the representational epistemology embraced by environmental crime science, this figure is discursively repressed.[14]

In terms of crime control, technology is consistently invested with agency and the power to reengineer sociopolitical relations in accordance with some image of the greater good. As we have had occasion to note, technology is fetishized. But the technologies that populate the field are diverse; they may entail organizing perception according to scientific taxonomies and philosophical logic or, perhaps more commonly, developing material devices. Beccaria and Lombroso embody the former tendency, evident in their claims that any given act or individual can be located within some broader category and that once this is done, the state can impose the "correct" punishment.[15] Following from its centering of the body, biological science fetishizes biometrics—DNA profiling and fingerprinting being the most obvious and well-known technologies. In actuarial science, it is the risk prediction instrument, which hinges on databases and the discovery of statistical correlations, that is most fetishized. And whereas security science places its faith in almost any technology that barricades space, environmental crime science is obsessed with spatial design and modification or novel devices that block the expression of those behaviors deemed undesirable.

The relationship between crime sciences and the operation of criminal justice is by no means simple. In some instances, crime science has direct implications and effects within criminal justice. It can lead to shifts in investigative practices, legislation, legal decision-making, and so on.[16] However, the line that links crime science and criminal justice need not always be obvious. It may be the case that crime sciences encourage very subtle changes in, for example, how legal concepts are to be understood and applied or what kinds of practices come to be seen as consistent with justice. Concerning some of the more obvious reverberations in criminal justice, bio-forensics has normalized the use of fingerprint and DNA technologies and their corollary practices; actuarialism has legitimized the use of risk assessment instruments in policing, sentencing, and parole, thereby encouraging a logic of preemptive decision-making and "incapacitation."

Alongside the more readily discernible effects of crime science are those reimaginings and transformations of criminal justice that occur somewhat beneath the radar, so to speak. In these respects, we have argued that bio-evidence—especially insofar as it purports to provide "scientific certainty"—does not scientize criminal justice so much as append different thresholds of truth to different crime types. That is, bio-evidence skews our sense of when and where the certainty of legal decisions ought to be desired and expected. And given the costs associated with biometric forms of evidence and their testing, such technologies effectively amount to a tool that favors prosecutors.[17]

Particularly evident in the areas of sentencing and parole, actuarial science inserts a logic of sacrifice into criminal justice. In rationalizing preemptive detention, for example, proponents of actuarialism posit that the state must choose between protecting nonoffenders (those who are "innocent," "worthy") or the rights of defendants (those who are "guilty," "unworthy").[18] However, preemptive decisions are better understood as entailing a choice concerning which kinds of "crime" are to be tolerated; what might be referred to as "false imprisonment" were it not orchestrated by the state is permitted because it is believed to prevent stereotypical street-level crimes.

The effects of security science are analogous. Alongside a privatization of state functions, here we see a framework of "criminal jurisprudence" being dislodged by "national security." The latter instantiates a "state of exception" in which the state grants itself the freedom to operate beyond constitutional rights and protections otherwise afforded citizens.[19] As various moments from the US-led war on terror aptly demonstrate, such as the detention and torture of "false positives," state atrocities have transpired under the banner of national security. Environmental science also seems to harbor similar effects, even if they are restricted to more regional contexts. Broken-windows policing, for instance, is anti-constitutional, but seems more focused on promoting police brutality within urban environments.[20]

In our critique of biosocial criminology, we noted that its proponents suffer from "science envy," or feelings of inadequacy that emerge when they compare their scholarly outputs to what they take to be "real" or "hard science." In no way would it be unfair to suggest that crime science is pervaded by such envy, the strange desire to satisfy the demands of a big daddy: science with a capital S. As a result, the discipline of criminology is often imagined as only able to secure its legitimacy by approximating positivist science. Criminology is thus encouraged to focus on "discovering" causal relations among "variables" and devising interventions that are "practical" or oriented toward the immediate, short-term resolution of problems. According to biosocial criminology, the criminologist can approximate "real science" by positioning "biology" as an independent variable in theorizing crime and drawing inspiration from the fields of genetics, evolutionary psychology, and neuroscience.[21] The body must be measured in some way. This is portrayed as core to the scientization of criminology, thereby ensuring its present and future relevance, even if it is—and this not without irony—indistinguishable from a return to Lombroso and thus reactionary.[22]

Actuarial, security, and environmental science appear much more preoccupied with reducing criminology to a technical specialization, a practical, "nuts and bolts" discipline that spends most of its energy developing gadgets and machines. For the actuarial, forecasting instruments and their refinement serves as the ultimate goal; for security, it is boundary fortifications and surveillance; and target-hardening technologies are the primary obsession for the environmental crime scientist. In these views, a criminology that resembles meteorology or engineering is desired.

No matter the variations on this theme of a scientific-technical criminology, it is quite consistent in its rejection of sociological criminology and postmodern or constructivist criminologies. In this, the images of criminology advanced within crime science prohibit critical thinking within the discipline. The existence of larger, structural problems may be acknowledged, but they are posited as beyond the purview of "scientific criminology," rendered too difficult or simply unnecessary for the technical specialist to address.[23] The most one can hope for is technological cover that may, at best, provide partial respite from the storm of sociopolitical disparities.

Despite its limits, we are skeptical that the authority—discursive and practical—of crime science will be dislodged any time soon. If anything, its position of prominence is intimately tied to its deficits. Rather than being a hindrance, its complicity with contemporary power asymmetries facilitates its popularity. Crime science may be riddled with contradictions, authoritarian, patronizing, and so on, but such problems are likely to be repressed insofar as the narratives and practices it proffers reinforce contemporary power relations. The racism and classism of environmental crime science, for example, do not engender a rejection of its logic and practice, but rather enhance its acceptance among much of the social body.

Of course, crime science will also be difficult to dislodge for what are somewhat more mundane, economic, and political reasons. The enterprise readily lends itself to commodification and is therefore a source of profit to some. The US-led war on terror has assured that billions of taxpayer dollars have been funneled into the coffers of private corporations, which supply the technologies necessitated by military aggression and enhanced security.[24] Although not quite occurring on the same scale, the technologies increasingly deployed to surveil everyday public life, especially as it is lived in the spaces of capital, also entail the transfer of public wealth into private hands.[25]

Interwoven with economic interests, crime and its control have risen to prominence as a central political issue in much of the Anglophone world. Rather than lively debate in which a range of possible approaches are entertained, however, one finds a punitive, monolithic hegemony. This is not because draconian penalties are effective. As has repeatedly been said of the prison, for example, it is more criminogenic than corrective.[26] And yet, while three of four US prisoners are rearrested after release from prison within five years, and although more than half of them wind up back in prison within five years, the political parties that purport to be fiscally conservative appear to have no qualms about spending $81 billion each year on corrections.[27] When the collateral costs of crime control are included (e.g., policing, courts, health care, and the like), the annual bill tops $181 billion.[28] Curiously, the fiscal hawks are among the loudest in calling for more laws, more police, and bigger prisons.

The "tough on crime" approach—increasingly repackaged as "smart on crime"—is so entrenched in contemporary political discourse that it is almost unthinkable for a politician to utter alternatives. To do so is political suicide. To defy public opinion—and thus be seen as championing the interests of predatory "criminals" who have forsaken their right to participate in society—is to alienate one's voter base. Thus, even when politicians believe that draconian crime legislation is misguided, the savvy ones know enough to "get out of the way" of fast-moving initiatives.[29] The lawmaker who calls for greater constitutional protections or a return to social welfare, or suggests interrogating any given use of the crime label, is seen as ignoring the public's rising fear and hatred of crime, thereby eradicating their chances of reelection. Even an effective public policy is politically useless unless it can be demonstrated to fix the crime problem within the span of a single election cycle. There is a reason that former Illinois governor George Ryan commuted the death sentences of all 167 condemned prisoners on death row just two days before leaving office in 2003. There is a reason that the district attorneys who write letters of support to ameliorate the damage of California's three-strikes law are usually *former* district attorneys.

We have hinted at alternatives throughout the text. In this respect, critical, sociological criminology has made an appearance, but it is a constructivist approach that has commanded most attention. Like crime science, critical criminology typically operates from within the confines of representational epistemology. It diverges, however, in how it theorizes criminalized behavior and thus embeds a different political agenda in some respects. For the critically inclined criminologist, criminalized behavior follows from social inequality, poverty, structural racism and discrimination, relative deprivation, and among other forces, socioculturally induced strain.[30] While such arguments often ontologize crime, they do move away from focusing on individuals and their supposed deficits. Such critical narratives posit that structural relations are the appropriate point of intervention: combating inequality and discrimination will curtail criminal motivations and thus control crime.

A major limit to this approach, however, is that it effectively operates as a heterodox position in which crime science typically provides the orthodox view, to appropriate Pierre Bourdieu's terminology.[31] Insofar as critical approaches remain a counterpoint to crime science, debate will be constrained by the perimeters of representational epistemology. Restricted in such a manner, one is often compelled to choose between causal theories of crime and then defend their choice by marshaling supporting facts and figures. Each position grants the other credence and reinforcement by entertaining it, while providing little opportunity to go beyond either viewpoint. Moreover, the doxa embedded in crime science secures advantages by default: crime is invariably positioned as a "dependent variable," it is presumed to correspond to some objective reality, it can be theorized, and it can be controlled.

The logic of constructivism disrupts this closed circle by offering an alternative to representational epistemology. Rather than assessing narratives in terms of how accurately or inaccurately they mirror reality, it regards them as constructing the worlds that they purport to map. Skeptics of this position can be found within critical criminologies and, not surprisingly, crime sciences. Opponents from both camps often claim that constructivism has little to offer insofar as it rejects reality.

Among those who endorse crime science, some claim that constructivism leaves crime problems intact.[32] The more critically inclined echo this sentiment but also note that dissolving the boundary between narrative and reality runs the risk of leaving power asymmetries undisturbed.[33] In these respects, both views construe constructivism as irresponsible and harmful.

Such critiques, however, seem to miss the point of constructivist logic. It is not so much that there is no reality per se, or that there is nothing outside the subject, but that we cannot access a pure, unmediated reality. The "real" is saturated with narratives, discourses, statements, cultural logics, and so on. Likewise, any analysis of reality is saturated in the same way. It is in this sense that narrative can be understood as an active force in sociopolitical relations and the crafting of cultural, epistemic frameworks. As we describe in chapter 1, discourse and practice are co-constituted, each always embedded in the other.

Rather than being dismissible as a rejection of reality or "semantics," constructivism ought to make us pause and ask about the effects that might follow from any given narrative.[34] What assumptions are embedded in and reproduced by a particular way of speaking about a problem? What practices and conceptual understandings are instantiated by specific discursive regimes? Answering such questions does not hinge upon performance of Donna Haraway's "god-trick"; one need not assume a vantage point from which all can be seen.[35] It does entail, however, recognizing that what may appear a neutral, objective portrayal of reality could quite easily be positioned as an "independent variable," a force that constructs the real. In this sense, constructivism involves a call for greater reflexivity, stepping back to consider the larger context of knowledge production.

In taking this step backward, one can discern that crime science—although it presents itself as a benevolent force, a purveyor of pragmatic solutions to crime problems—is replete with adverse effects. Very rarely, for example, does it stop to interrogate when and how the label of "crime" is applied. Incidentally, neither does it question avoidance of the crime label. And, as Nicolas Carrier notes, it does not seem to recognize that the distinction between criminal and noncriminal, illegal and legal, is thoroughly bound to the exercise of state power.[36]

The problematic consequences of this presumptuousness are evident throughout crime science, some of which we have attempted to chart throughout this book. Biological crime science accepts the notion that some behaviors can be classed as *mala in se* crimes—or crimes that are inherently evil—and then uses this category to bio-pathologize Black people. However, the category is abandoned the very moment it must be conceded that dominant, white groups have a long history of committing such crimes. Quite suddenly, alternative labels are found and utilized when it is a matter of describing the actions of dominant social groups. Genocide and slavery, for example, are rendered "odious" or "malfeasant," but they do not embody evil in itself and are thus not considered *mala in se* crimes.[37] This simultaneously promotes and justifies a racialized use of state power.

Moreover, bio-forensic evidence associates different standards of truth with different offending types. Legal systems may have internalized, and now expect to see, scientific notions of truth being met or surpassed when it comes to adjudicating cases involving murder, particular forms of sexual violence, and some property crimes. But this is not necessarily the case throughout any given legal system. As evidenced by Corinna Kruse's study, the "scientization" of criminal justice can lead to the downplaying and minimizing of erroneous findings of guilt and thus the injustice of imposing unwarranted punishments. This is especially likely to occur in relation to forms of offending deemed "minor," but which happen to saturate criminal justice systems.[38] In this, bio-forensic technologies encourage a cavalier attitude concerning the freedom of those othered by criminal justice systems. Likewise, actuarial science effectively embraces a reckless stance when it comes to the freedom of those who find themselves ensnared by its prediction instruments. This is evidenced by its varied efforts to neutralize the "false positive" dilemma by which it is plagued.[39] Security and environmental sciences are perfectly happy to burden other territories and their agents of control with whatever crimes or threats they seek to block. Moreover, both seem more than willing to privatize punitive forms of control and abandon human rights by promising to deliver security to some portion of the population.

Broadly speaking, crime science intensifies the social marginalization of those from already marginalized social spaces, thereby extending that proportion of the population for whom, in all likelihood, absorption into society appears highly unlikely. Many of crime science's particular strategies seek to control crime by providing retrospective evidence that helps secure convictions (e.g., biometrics, CCTV) or by incapacitating individuals (e.g., risk assessment in sentencing contexts). Such strategies assume that crime control can be reduced to punishment. The investment that crime science makes in punitive control strategies ensures the permanent exclusion of those who become entangled in its machinations. It is a logic in which "carrots" are simply excluded or transformed into "sticks."

There is a need, then, to interrogate the epistemological assumptions embedded in crime science. Arguably, the primary assumption of crime science is that there exists an independent, external reality that readily lends itself to unfiltered observation and theorization. From here, it is further presumed that political, normative judgments automatically follow from this supposed capacity to record reality. Rather than accept such tenets, it is important to work from the premise that political standpoints inform what is construed as a problem and how it is approached. Such standpoints also inform any leap that is made from representation to moral, normative evaluation.

To elaborate this latter point, it should be somewhat obvious that a moralistic lens, a judgment concerning what is "right" and "wrong," "orderly" or "disorderly," informs how and when behaviors will be coded as "criminal." Pegged to such right-wrong demarcations is demand for the punishment—often excessive—of infractions. In the case of illicit drug use—a practice that is very much connected to the problem of "mass incarceration" in places like the United States and the United Kingdom—it seems clear that moral sentiment has triumphed to ensure that what could just as easily be coded as a collective *social* (or public health) issue continues to be framed as an individual *criminal* problem.[40] The effect of this is to minimize, if not eradicate, support for those who might seek it and who might benefit from it. Addicts without access to safe drugs resort to illegal supply, risking violent situations, arrest, and unregulated products. Overdoses and the spread of HIV and hepatitis C are predictable consequences. So is state reliance on incarceration: one in five US prisoners is incarcerated for drugs. And when these people come out of prison, they are often ineligible for social safety net benefits such as welfare, food stamps, public housing, and student financial aid. This may satisfy the desire for vengeance and the imposition of order that some social groups seem to have, but whether it amounts to sound social policy is questionable.[41] A not-so-morally governed set of policies might entail extracting drug use from criminal justice institutions and relocating it within rehabilitation and counseling settings. This would presumably entail harm reduction, decriminalization, and responsible supply as well as imagining new social spaces capable of addressing drug dependency.

According to the constructivist position, narratives cannot be exempt from scrutiny for their performative effects, a point that applies no matter how objective any particular narrative may seem. Concerning the discourses proffered by crime science, there is a need to continue questioning its language of "variables," which allows the scholar's conceptual choices and preferred categories to masquerade as nothing more than mirrors of the real. Rather than "crime" being prescribed by reality, for example, crime science selects the construct from a range of possibilities.[42] Along similar lines, it remains important to recognize and call into question the arbitrary nature of how variables are organized in crime science and its associated criminological theories. It is not difficult to sense that the constructs typically positioned as "independent variables" could readily be reframed as "dependent variables," or vice versa.[43] Rather than aspects of physical appearance explaining criminality, for example, it may well be that theories are attributing criminality to physical features. In short, instead of variables simply encoding their status as either "cause" or "effect"—as if these were akin to the EXIF data tethered to digital images—it is theory that fashions variables into an arrangement.

By interrogating these types of assumptions, the politics and normative prejudices embedded in knowledge production can be brought to the surface and revealed. For the most part, positioning "crime" as an effect implies that its antecedents constitute points of intervention. Because crime science construes defective uses of rationality or biological dispositions as the fundamental cause of crime—and often dismisses sociocultural forces and possibilities—it is invariably individuals and social groups that constitute the point of anti-crime interventions. In promising that crime will ultimately be brought under control, crime science asserts that it cannot be anything other than a benevolent social force, a project that seeks to address what common sense dictates is undesirable. Crime science never seems to entertain the possibility that it constitutes a problem, that it amounts to an exercise of power. Unraveling this delusion, however, is among the tasks of constructivist criminology.

Notes

INTRODUCTION

1. Doyle, "Boscombe Valley Mystery," 124.

2. Berg, "Sherlock Holmes."

3. Clarke, "Crime Science," 271. The rest of this paragraph is derived from table 14.2 in Clarke's chapter.

4. We should emphasize that our critique of crime science is based on an analysis of texts and practices associated with Anglophone countries, especially the United States, the United Kingdom, Canada, Australia, and New Zealand. This is a limitation of our approach: it cannot be assumed that the Anglophone world illustrates a comprehensive universe of crime science; other places evince diverse logics and technological practice. Nevertheless, insofar as crime science amounts to a discourse, it is free floating and thus has the potential to pervade any political context, thereby exerting its performative effects.

5. Norris, "Success of Failure"; Sivarajasingam et al., "Effect of Urban"; Jacobs, "Manipulation of Fear"; Jacobs et al., "Carjacking"; Harcourt, *Against Prediction* (2007).

6. Collins, "And the Walls"; Eterno and Silverman, "New York City"; Gras, "Legal Regulation"; Howell, "Broken Lives"; Lippert and Wilkinson, "Capturing Crime," 140; Wilson and Sutton, "Watched Over."

7. Werth, "Theorizing the Performative."

8. Concerning sociological narratives, see Hall et al., *Criminal Identities*; Merton, "Social Structure"; Miller, "Lower-Class Culture.". Concerning constructivist narratives, see Carrier, "Academics' Criminals"; Carrier, "Critical Criminology"; Hulsman, "Critical Criminology."

9. See, for example, Clarke, *Situational Crime.*

10. Gerlach, *Genetic Imaginary*; Quan, "Black and White."

11. Martin, "Bombs, Bodies."

12. Carrier, "Speech for the Defense"; Harding, *Science Question*; Haraway, "Situated Knowledges"; Henry and Milovanovic, *Constitutive Criminology.*

13. E.g., Barlow, *Introduction to Criminology*, 479, 541; Gottfredson, *Exploring Criminal Justice*, 72–73; Paternoster and Bachman, *Explaining Criminals*, 13, 48.

14. The terms "bio-forensics" and "bio-evidence" are intended to capture forms of forensic evidence that are a product of, or directly tied to, the body. DNA and fingerprints are the most obvious examples. The terms are used throughout to distinguish biological forms of evidence from other types of forensic analysis, such as tool mark analysis and handwriting comparisons.

15. Cole, *Suspect Identities*; Cole, "More Than Zero."

16. Aronson and Cole, "Science and the Death Penalty"; Lynch et al., *Truth Machine.*

17. Quinlan, "Technoscience."

18. Gerlach, *Genetic Imaginary* ("scientize"); and Kruse, *Social Life* (tolerable).

19. Walsh, *Race and Crime*, 66.

20. Walsh, *Race and Crime*, 65.

21. Harcourt, *Against Prediction.*

22. Feeley and Simon, "New Penology"; Feeley and Simon, "Actuarial Justice"; Garland, *Culture of Control* (punitive); and O'Malley, *Crime and Risk* ("soul of the offender").

23. Bueno, "Face Revisited"; Eaglin, "Against Neorehabilitation"; Hannah-Moffat, "Criminogenic Needs"; Werth, "Individualizing Risk"; Werth, "Risk and Punishment."

24. Agamben, *Home Sacer.*

25. On these arguments, see Meehl, *Clinical versus Statistical*; Gottfredson and Gottfredson, "Selective Incapacitation?"; Berk, "Balancing the Costs".

26. Henne and Shah, "Unveiling White Logic"; Henne and Troshynski, "Intersectional Criminologies"; Van Cleve and Mayes, "Criminal Justice."

27. Roussell and Gascón, "Defining 'Policeability.'".

28. Klein, *Shock Doctrine* (2007).

29. Harvey, "Fetish of Technology."

30. Balkin, "Constitution"; Conrad, "Executive Order"; Jordan, "Decrypting"; Katyal and Caplan, "Surprisingly Stronger"; Loader, "Thinking Normatively."

31. Foucault, *Discipline and Punish.*

32. Harvey, *Brief History.*

33. Kelling, "Crime Control"; Wilson and Kelling, "Broken Windows."

34. Fagan and Davies, "Street Stops."

35. See especially Kelling, "Crime Control."

36. Eterno and Silverman, "New York City"; Harcourt and Ludwig, "Reefer Madness"; Howell, "Broken Lives"; Moore, "Sizing up Compstat"; Walsh, "Compstat."

37. Clarke, *Situational Crime.*

38. Walsh and Beaver, "Introduction." See also Rafter, *Criminal Brain.*

39. Walsh, *Race and Crime*; Walsh and Wright, "Rage against Reason."
40. Carrier, "Academics' Criminals."

CHAPTER 1. A BRIEF SKETCH OF CRIME SCIENCE

1. Cf. Berger and Luckmann, *Social Construction of Reality*.
2. Foucault, *Archaeology*; Foucault, *Discipline and Punish*; Foucault, *Power/Knowledge*.
3. We chart these positions through the work of Nicolas Carrier (radical constructivism), Sandra Harding (successor sciences), Donna Haraway (situated knowledges), and Henry and Milovanovic (constitutive criminology).
4. Beccaria, *On Crimes*; Lombroso, *Criminal Man*.
5. On this point, see Horn, *Criminal Body*, 146, who warns against dismissing crime science as "pseudo-science" and thereby failing to appreciate its practical impacts irrespective of its limits.
6. Foucault, *Archaeology*; Foucault, *Power/Knowledge*.
7. Marx, "German Ideology"; Marx, "Preface"; Marx, "Communist Manifesto."
8. Carrier and Walby, "Ptolemizing Lombroso," 20; Carrier, "Speech for the Defense," 173. The formulation quoted is derived from Niklas Luhmann, *Theories of Distinction*. This distinction corresponds broadly to the difference between phenomena (observable qualities) and noumena (things-in-themselves) in Kantian philosophy. Kant, *Critique of Pure Reason*.
9. Carrier, "Academics' Criminals," 10.
10. Carrier, "Speech for the Defense," 176.
11. Nietzsche, *Will to Power*.
12. Haraway, "Situated Knowledges."
13. Carrier, "Speech for the Defense," 173. Carrier illustrates the point as follows: "'X' may be constructed as 'criminal' by the juridical system or various activists, but, for other observers, it may be 'pleasure', a 'symptom', an 'economic activity', a 'conflicting situation', and so on. *To establish that 'x' really is* 'criminal' (or a 'symptom', etc.) is a matter of social influence, power, or force."
14. Carrier and Walby, "Ptolemizing Lombroso." See also Walby and Carrier, "Rise of Biocriminology."
15. Carrier and Walby, "Ptolemizing Lombroso," 7.
16. Carrier and Walby, "Ptolemizing Lombroso"; see also Walby and Carrier, "Rise of Biocriminology."
17. Carrier and Walby, "Ptolemizing Lombroso," 1–2.
18. Walby and Carrier, "Rise of Biocriminology," 275.
19. Carrier and Walby, "Ptolemizing Lombroso," 18. See also Walby and Carrier, "Rise of Biocriminology," 274.
20. Carrier and Walby, "Ptolemizing Lombroso," 21. See also Walby and Carrier, "Rise of Biocriminology."
21. Carrier and Walby, "Ptolemizing Lombroso," 21.
22. See also Carrier, "Critical Criminology," 337, 341.
23. Harding, *Science Question*; Harding, *Objectivity and Diversity*.
24. Harding, *Objectivity and Diversity*, 47. See also Kuhn, *Structure*.

25. Harding, *Science Question*, 35–40, 56. See also Harding, "Introduction," 16.
26. Harding, *Objectivity and Diversity*, 30; Harding, *Science Question*, 100.
27. Harding, *Science Question*; esp. 238–39, 249–251.
28. Harding, *Science Question*, 58–81.
29. Harding, *Science Question*, 58–81.
30. Harding, *Objectivity and Diversity*.
31. Harding, *Objectivity and Diversity*, 126–29. See also Harding, *Science Question*, 249–51.
32. Haraway, "Situated Knowledges," 578. Cf. Harding, *Objectivity and Diversity*, 159.
33. Haraway, "Situated Knowledges," 578.
34. Haraway, "Situated Knowledges," 579.
35. Haraway, "Situated Knowledges," 581.
36. Haraway, "Situated Knowledges," 584–90 ("see better"); this is Haraway's reading of Harding. See also Harding, *Objectivity and Diversity*, 30–31, 42, 161. Haraway, "Situated Knowledges," 586 ("perfect subject of oppositional history").
37. Haraway, "Situated Knowledges," 586.
38. Haraway, "Situated Knowledges"; the quoted material appears on page 590.
39. Henry and Milovanovic, *Constitutive Criminology*.
40. Henry and Milovanovic, *Constitutive Criminology*, 6. The notion appears throughout the text.
41. See the pieces by Carrier (and colleagues) cited previously, especially Carrier, "Academics' Criminals," in which the attempt to find the causes of crime is framed as emblematic of the pathologizing gaze.
42. Carrier, "Academics' Criminals," 103.
43. Henry and Milovanovic, *Constitutive Criminology*, 105.
44. Henry and Milovanovic, *Constitutive Criminology*, 106.
45. For a detailed account that calls the ontological status of crime into question, see Hulsman, "Critical Criminology."
46. Henry and Milovanovic, *Constitutive Criminology*, 111.
47. Henry and Milovanovic, *Constitutive Criminology*, 112.
48. Henry and Milovanovic, *Constitutive Criminology*, 116 (emphasis in original).
49. See also Carrier "Speech for the Defense," 172.
50. Henry and Milovanovic, *Constitutive Criminology*, 118.
51. Henry and Milovanovic, *Constitutive Criminology*, 119.
52. Mills, *Sociological Imagination*, 77–78.
53. Henry and Milovanovic, *Constitutive Criminology*, 115.
54. Carrier, "Academics' Criminals," 12.
55. Carrier, "Academics' Criminals."
56. Harding, *Science Question*, 245; Henry and Milovanovic, *Constitutive Criminology*, 5.
57. Carrier, "Critical Criminology," 342. See also Carrier, "Speech for the Defense," 180, in which the political nature of radical constructivism is difficult to deny.

58. For an unequivocal statement along these lines, see Harding, *Science Question*, 124–25. The idea suffuses the other epistemological positions mentioned.

59. E.g., Barlow, *Introduction to criminology*, 479, 541; Gottfredson, *Exploring Criminal Justice*, 72–73; Paternoster and Bachman, *Explaining Criminals*, 13, 48. See also Gibson, *Born to Crime.*

60. Carrier, "Speech for the Defense."

61. Lindesmith and Levin, "Lombrosian Myth."

62. Beccaria, *On Crimes*, 11.

63. Lombroso, *Criminal Man.*

64. Lombroso, *Criminal Man.* We return to the "born criminal" and Lombroso's typology of criminals later in the chapter.

65. Lombroso, *Criminal Man*, 171. See also Evans, *Criminal Prosecution.*

66. Beccaria, *On Crimes*, 43, 47.

67. Lombroso, *Criminal Man*, 165; Ferrero, *Criminal Man.*

68. Beccaria, *On Crimes*, 15.

69. Beccaria, *On Crimes*, 15.

70. Beccaria, *On Crimes*, 63, 99.

71. CNN, "Roberts: My Job Is to Call Balls."

72. Lombroso, *Criminal Man*, 268.

73. Lombroso, *Criminal Man*, 91.

74. Becker and Wetzell, *Criminals and Their Scientists.*

75. Lombroso, *Criminal Man*, 221, 253.

76. Lombroso, *Criminal Man*, 348.

77. Lombroso, *Criminal Man*, 346.

78. Cf. Samenow, *Myth of the Out of Character Crime.*

79. Beccaria, *On Crimes*, 93.

80. See also Bentham, *Introduction to the Principles.*

81. Lombroso, *Criminal Man*, 197.

82. Lombroso, *Criminal Man*, 348.

83. Lombroso, *Criminal Man*, 342.

84. Lombroso, *Criminal Man*, 330–54.

85. See Rafter, "Criminology's Darkest Hour."

86. Lombroso, *Criminal Man*, 197.

87. Lombroso, *Criminal Man*, 313.

88. Beccaria, *On Crimes*, 17.

89. Oleson, "Blowing Out All the Candles," 708.

90. Beccaria, *On Crimes*, 16. It may be worth noting that inconsistent punishment is by no means a problem of the past. For example, studies still find harsher penalties are handed down to those from racially marginalized groups, and this despite the introduction of sentencing guidelines intended to combat discriminatory sentencing practices (Steffensmeier et al., "Interaction of Race, Gender, and Age"; Shermer and Johnson, "Criminal Prosecutions"; Spohn, "Effects of the Offender's Race"). If anything, inconsistency in punishment due to sociological factors remains a dominant theme in sentencing studies.

91. Berman and Feinblatt, *Good Courts*. See also Kramer, "Neoliberal States"; Kramer, "Differential Punishment."

92. Horn, *Criminal Body*, 145–46.

93. Rafter, "Criminology's Darkest Hour" (2008).

94. Tonry, "Ethnicity, Crime," 12.

95. Alexander, *New Jim Crow*; Cunneen, "Racism, Discrimination"; Davis, *Are Prisons Obsolete?*; McIntosh and Coster, "Indigenous Insider Knowledge"; Wacquant, *Prisons of Poverty*. For a critique of "mass incarceration" see Carrier, "Anglo-Saxon Sociologies," which argues that "mass incarceration" presupposes that an appropriate level of incarceration can somehow be determined. This is quite problematic from a prison abolitionist perspective. See also Simon, "Mass Incarceration."

96. Oleson, "Risk in Sentencing" (2011) (sentencing process); and Martinson, "What Works?" (1974) (harsher punishments are necessary). See also Zimring and Hawkins, *Scale of Imprisonment*, 89: "Recognition that imprisonment's distinctive feature as a penal method is its incapacitative effect has implications for criminal justice policy. The contribution of prisons to crime control by way of rehabilitation programs or individual and general deterrence is problematic. But there can be no doubt that an offender cannot commit crimes in the general community while incarcerated."

97. For an interesting argument that ties physiognomy to social capital, see Mocan and Tekin, "Ugly Criminals."

98. Horn, *Criminal Body*, 131, 145 (identify sex offenders). See Raine, "Annotation"; Raine, *Anatomy of Violence*; and Raine et al., "Increased Executive Functioning" (brain scans of psychopaths). See Horn, *Criminal Body*, 145 ("'decode' and 'read' the human genome").

99. Horn, *Criminal Body*, 146.

100. For example, Samenow, *Inside the Criminal Mind*. See also Kramer et al., "Neoliberal Prisons."

101. Packer, *Limits of the Criminal Sanction*, 74–75.

102. Tonry and Lynch, "Intermediate Sanctions."

103. Foucault, *Discipline and Punish*, 75–82.

104. Lombroso, *Criminal Man*, 92.

105. Lombroso notes the critique in *Criminal Man*, 163–65.

106. Lombroso, *Criminal Man*, 165.

107. See Walby and Carrier, "Rise of Biocriminology," 272–74.

108. Harding, *Science Question*, 77.

109. Harding, *Science Question*, 251. See also Harding, "Introduction," 15.

110. Carrier, "Speech for the Defense." The quotes are from pages 179 and 180, respectively.

CHAPTER 2. BIOLOGICAL CRIME SCIENCE

1. Quinlan, "Technoscience," 333. Quinlan is drawing from Haraway, *Modest_Witness*.

2. Cole, *Suspect Identities*; Lynch et al., *Truth Machine*. Enhancing penalties for recidivists is so commonplace that to argue *against* penalty enhancement for criminal history is novel, even radical. See Frase and Roberts, *Paying for the Past*.

3. The quotes from Brayley and Galton are from Cole, *Suspect Identities*, 190–91 and 99, respectively.

4. On this point, see Gerlach, *Genetic Imaginary*, 11.
5. Kruse, *Social Life*.
6. Gerlach, *Genetic Imaginary*, 35.
7. Gerlach, *Genetic Imaginary*, 8, 87–89. See also Rosen, "Liberty, Privacy."
8. Kruse, *Social Life*.
9. Knorr Cetina, *Epistemic Cultures*.
10. Kruse, *Social Life*, 145–46.
11. Gerlach, *Genetic Imaginary*, 165.
12. Gerlach, *Genetic Imaginary*, 166.
13. Foucault, *Discipline and Punish*.
14. Gerlach, *Genetic Imaginary*, 68, 79. It is worth noting that Gerlach's notion of postpanoptic risk management resonates with Feeley and Simon's "new penology" argument. See Feeley and Simon, "New Penology."
15. Doe quoted in Quinlan, "Technoscience," 333 (emphasis added).
16. Katz, *Seductions*.
17. Kruse, *Social Life*, 104.
18. Kruse, *Social Life*, 124.
19. Gerlach, *Genetic Imaginary*, 195. On eluding bio-technoscience, see also Rosen, "Liberty, Privacy," 41; Simoncelli, "Dangerous Excursions," 393.
20. Carrier and Walby, "Ptolemizing Lombroso." See also Walby and Carrier, "Rise of Biocriminology."
21. Gerlach, *Genetic Imaginary*, 192.
22. Cole, *Suspect Identities*, 100.
23. Cole, *Suspect Identities*, 100.
24. Harcourt, *Against Prediction*. For a comparable argument, see also Gerlach, *Genetic Imaginary*, 166.
25. Cole, *Suspect Identities*, 63–65, 96.
26. Cole, *Suspect Identities*, 258.
27. The quoted material is from Lynch et al., *Truth Machine*, 151 (citations omitted).
28. Select Committee on Home Affairs, "Nature and Extent."
29. For the database figures, see Home Office, "Official Statistics."
30. Kruse, *Social Life*, 124–30; Gerlach, *Genetic Imaginary*, 91, 134. Both note that DNA analysis is closely aligned with the state, thereby compromising the possibility of objectivity.
31. Lynch et al., *Truth Machine*, 235. This is not to mention the problem that defense attorneys may encounter if forensic evidence is analyzed with patented technologies, which can make some lines of cross-examination next to impossible.
32. Gerlach, *Genetic Imaginary*, 171.
33. Quan, "Black and White," 1412; Matejik, "DNA Sampling."
34. Quan, "Black and White," 1411.
35. Quan, "Black and White," 1442.
36. Quan, "Black and White," 1412.
37. Quan, "Black and White," 1433.
38. Quan, "Black and White," 1430. See also Duster, "Selective Arrests." The imposing of racial groups upon a continuum of genetic variation is discernible in

biosocial criminology, especially in its efforts to assert that "race is real." We will touch upon the conflation of ancestry and race, albeit briefly, when we discuss biosocial criminology below.

39. Gerlach, *Genetic Imaginary*, 192.

40. Lynch et al., *Truth Machine*, 339.

41. Cole, *Suspect Identities*, 291; Lynch et al., *Truth Machine*, 29; Gerlach, *Genetic Imaginary*, 184.

42. Cole, *Suspect Identities*, 292.

43. Gerlach, *Genetic Imaginary*, 184–85. The second dragnet involves sexual assault and murder, the third a violent murder.

44. The quoted material in this paragraph is taken from Gerlach, *Genetic Imaginary*, 185.

45. Surette, *Media, Crime*, 204.

46. Aronson and Cole, "Science and the Death Penalty."

47. See Gerlach, *Genetic Imaginary*, 98–132.

48. Latour, *Science in Action*.

49. Gerlach, *Genetic Imaginary*, 144–52; Lynch et al., *Truth Machine*, 228–55.

50. For critiques of fingerprinting, see Cole, *Suspect Identities*; Cole, "More Than Zero"; "'Opinionization'"; Dror and Mnookin, "Use of Technology"; Zabell, "Fingerprint Evidence"; Giannelli, "Daubert Challenges," 630; Grieve, "Possession," 521–23; Mnookin, "Fingerprint Evidence." For extended critiques of DNA typing, see Lynch et al., *Truth Machine*, 86, 111; Gerlach, *Genetic Imaginary*; Kruse, *Social Life*.

51. Cole, *Suspect Identities*, 261.

52. Gerlach, *Genetic Imaginary*, 173.

53. Gerlach, *Genetic Imaginary*, 174. The notion of scientization is recurrent; see also 48, 66, 217.

54. Lynch et al., *Truth Machine*, 257, 306. See also see Cole, *Suspect Identities*, 302; Gerlach, *Genetic Imaginary*, 30, 129, 200–204. Cf. Loftus and Palmer, "Reconstruction" on the weakness of eyewitness testimony.

55. Cole, *Suspect Identities*, 293.

56. Lynch et al., *Truth Machine*, 339.

57. Kruse, *Social Life*, 135.

58. Carrier and Walby, "Ptolemizing Lombroso"; Walby and Carrier, "Rise of Biocriminology."

59. Walsh, *Race and Crime*.

60. Walsh, *Race and Crime*, 1–37. See also Wright, "Inconvenient Truths," 139–43.

61. See for example Walsh and Wright, "Rage against Reason," 68; Wright, "Inconvenient Truths."

62. As Walsh says in *Race and Crime*, "Substituting *ethnic* for *racial* does nothing but replace one term with another" (p. 4). See also Walsh and Yun, "Race and Criminology," 1281.

63. Carrier and Walby, "Ptolemizing Lombroso," 21.

64. Terms such as "African American" and "Asians" are repeatedly used by Walsh (and other biosocial criminologists). We retain the use of such terminology when summarizing or representing their arguments in order to preserve the political sensibility that informs the discourse of biosocial criminology.

65. Walsh and Beaver, "Introduction," 9.

66. We return to some of the problems with this biosocial model in the following section.

67. See, for example, Wright, "Inconvenient Truths," 138, 151.

68. Walsh, *Race and Crime*, 20. See also Walsh and Wright, "Rage against Reason," 64.

69. Walsh, *Race and Crime*, 56.

70. Walsh, *Race and Crime*, 65 ("most odious practice") and 62 ("horrific").

71. Walsh, *Race and Crime*, 62.

72. Walsh, *Race and Crime*, 20.

73. For the "warrior gene" and MAO-A, see Brunner et al., "Abnormal Behavior"; Gibbons "Tracking"; Holland and DeLisi, "Warrior Gene"; Lea and Chambers, "Monoamine Oxidase." For critiques, see Merriman and Cameron, "Risk-Taking"; Hook "Warrior Genes," 5.

74. Walsh, *Race and Crime*, 56. The problem of "adaptive adaptations" and "maladaptive adaptations" within biosocial criminology is pursued later.

75. The quoted material is taken from Walsh, *Race and Crime*, 62–63 and 89.

76. Walsh, *Race and Crime*, 89.

77. Walsh, *Race and Crime*, 70.

78. Walsh, *Race and Crime*, 92.

79. Rafter, *Criminal Brain*, 125; Rafter, *Creating Born Criminals.*

80. Walsh, *Race and Crime*, viii.

81. Walsh, *Race and Crime*, 20.

82. Walsh, *Race and Crime*, 20. For the same refrain, see also Wright, "Inconvenient Truths," 138.

83. Thornhill and Palmer, *Natural History of Rape* (2000) ("evolutionary" explanation). See also the discussion in Rafter, *Criminal Brain*, 211.

84. Walsh, *Race and Crime*, 4. The same sentiment recurs in Walsh and Yun, "Race and Criminology," 1289.

85. Rafter, *Criminal Brain*, 182–86.

86. Rafter, *Criminal Brain*, 191.

87. On the link between science, society, and power, see also Henne and Troshynski, "Intersectional Criminologies."

88. Rocque et al., "Biosocial Criminology," 306.

89. Walsh and Wright, "Rage against Reason," 69.

90. On naturalizing rape, see Walsh and Beaver, "Introduction," 18; Quinsey, "Evolutionary Theory"; Walsh, "Evolutionary Psychology"; Thornhill and Palmer, *Natural History of Rape.*

91. Walsh and Beaver "Introduction," 7.

92. Walsh and Yun "Race and Criminology," 1290; Wright, "Inconvenient Truths," 143–44.

93. Walsh, *Race and Crime*, 23. For the uncritical use of figures, see also Wright, "Inconvenient Truths," 144; Walsh, "Crazy by Design," 154.

94. For discussion of the limits of official statistics, see Junger-Tas and Marshall, "Self-Report Methodology" (1999); Kitsuse and Cicourel, "Note on the Use of Official Statistics"; Oleson, *Criminal Genius.*

95. Walsh, *Race and Crime*, 47 (citations omitted; emphasis in original).

96. Walsh, *Race and Crime*, 48.

97. Walsh, *Race and Crime*, 47.

98. Walsh, *Race and Crime*, 48.

99. For a further example of this tendency, see Wright, "Inconvenient Truths," 144.

100. Baker et al., "Genetic and environmental"; Farrington and Loeber, "Epidemiology"; Jaffee et al., "Nature x Nurture"; Raine, "Annotation"; Rhee and Waldman, "Genetic and Environmental," 497; Rowe, "Biometrical Genetic Models"; Tuvblad et al., "Heritability."

101. Cartwright, *Evolution*; Rowe, "Adaptive Strategy"; Walsh and Beaver, "Introduction," 9; Walsh, "Crazy by Design," 158.

102. Darwin, *Origin*.

103. For biosocial criminology's debt to Darwin, see Walsh and Beaver, "Introduction," 13–14.

104. A point that seems to be acknowledged within biosocial criminology. See Walsh and Beaver, "Introduction," 14.

105. For this version of maladaptation, see Walsh and Beaver, "Introduction," 14.

106. Rowe, *Biology and Crime* (2002); Walsh and Beaver, "Introduction," 17.

107. Walsh, *Race and Crime*, 118.

108. Walsh, *Race and Crime*, 119.

109. Walsh, *Race and Crime*, 126.

110. The formulation owes an obvious debt to Douglas, *Purity and Danger*.

111. The eugenic tendencies in biosocial criminology are evident in its subtle suggestions of resolving crime via drug therapies, the imposition of adherence to normative sexual practices, "early" interventions, and so on (see Walsh, "Crazy by Design," 163; Ghiglieri, *Dark Side of Man*). It may be worth noting that Gerlach describes a new kind of eugenics being derived from contemporary biotechnologies and the "genetic imaginary" with which they are associated. Rather than eugenics operating in a "top down" fashion, as has historically been the case, Gerlach suggests that its resurgence will most likely take infra forms, transpiring due to countless microlevel choices to manipulate biology (see Gerlach, *Genetic Imaginary*, 29). Cole (*Suspect Identities*, 305) also notes that biological theories of criminality—often stimulated by identification techniques—cannot avoid eugenic implications. It is almost as though the biosocial criminologist is alone in not seeing the eugenic implications of "biosocial" arguments. See also Duster, *Backdoor to Eugenics*; Oleson, "New Eugenics."

CHAPTER 3. ACTUARIAL SCIENCE

1. Harcourt, *Against Prediction*. We should emphasize that actuarial science has a long history. Recently, however, its incursion upon criminal justice owes a significant debt to a much more encompassing obsession with risk, its assessment, and its control. Although we acknowledge the importance of "risk society" type arguments, and although we occasionally draw from literature written in that vein, this chapter retains a focus on actuarialism and criminal justice.

2. See Oleson, "Risk in Sentencing."

3. Wilson and McCulloch, *Pre-crime.*

4. Costanza-Chock, "Design Justice."

5. Henne and Shah, "Unveiling White Logic"; Van Cleve and Mayes, "Criminal Justice".

6. See, for example, Robinson, "Punishing Dangerousness"; Morse, "Blame and Danger"; von Hirsch, "Selective Incapacitation Re-examined."

7. See Morris, "Persons and Punishment."

8. Feeley and Simon, "New Penology"; Feeley and Simon, "Actuarial Justice"; Garland, *Culture of Control.*

9. Feeley and Simon, "New Penology"; Feeley and Simon, "Actuarial Justice."

10. Feeley and Simon, "Actuarial Justice," 178.

11. Feeley and Simon, "New Penology," 453.

12. Feeley and Simon, "Actuarial Justice," 181–83.

13. Feeley and Simon, "New Penology," 455, 465.

14. Feeley and Simon, "New Penology," 455; Feeley and Simon, "Actuarial Justice," 179.

15. Garland, *Culture of Control*, 79.

16. Garland, *Culture of Control*, 82–90.

17. Garland, *Culture of Control*, 110.

18. For these three discourses, see Garland, *Culture of Control*, 61–63, 131, 184, and 129, respectively. We return to "situational crime prevention" in chapter 5.

19. Garland, *Culture of Control*, 140–44, 167–92.

20. Foucault, *Discipline and Punish.*

21. On this tendency, see also Blackmore and Welsh, "Selective Incapacitation," 520.

22. O'Malley, *Crime and Risk*, 11–13.

23. O'Malley, *Crime and Risk*, 19. The notion that risk is devoid of any political essence animates much of the work by O'Malley and colleagues. See, for example, O'Malley, "Risk, Power"; Smith and O'Malley, "Driving Politics."

24. O'Malley, *Crime and Risk*, 84. The notion of "reservation" is from Deleuze, "Postscript." See also Deleuze and Guattari, *Thousand Plateaus*, 358–60.

25. Wacquant, "Crafting the Neoliberal State."

26. Wacquant, "Crafting the Neoliberal State," 206.

27. Concerning the idea that risk is not plastic, malleable, or open to interpretation because it is always embedded in prevailing political ideologies, see also Mason and Magnet, "Surveillance Studies," 116; Feeley and Simon, "Actuarial Justice," 189–93.

28. Werth, "Individualizing Risk."

29. Werth, "Theorizing the Performative"; "Risk and Punishment."

30. Werth, "Theorizing the Performative," 343.

31. Hannah-Moffat, "Losing Ground." See also Hannah-Moffat, "Criminogenic Needs." It is worth noting that advocates often acknowledge that risk assessment and individualized discipline have fused but, unsurprisingly, construe such mergers as evidence of growing technological sophistication in the predictive arena. See, for example, Andrews et al., *Level of Service*; Bonta and Andrews, "Risk-Need-Responsivity"; Dvoskin and Heilbrun, "Risk Assessment."

32. Hannah-Moffat, "Losing Ground," 379.
33. Bueno, "Face Revisited," 73.
34. Bueno, "Face Revisited," 84.
35. Both quotes are from Bueno, "Face ervisited," 84.
36. Bueno, "Face Revisited," 83.
37. Bueno, "Face Revisited," 87.
38. Bueno, "Face Revisited," 81.
39. It is worth recalling that the continued presence of the individual is acknowledged by Feeley and Simon in "Actuarial Justice."
40. Kierkegaard, *Fear and Trembling*.
41. Kierkegaard, *Fear and Trembling*, 35, 48–52, 64–71.
42. Agamben, *Homo Sacer*.
43. Agamben, *Homo Sacer*. It might be worthwhile to draw attention to the work of Dale Spencer, who utilizes Agamben to theorize sex offender legislation. See Spencer, "Sex Offender as Homo Sacer."
44. Agamben, *Homo Sacer*, 115, 119–25.
45. Floud, "Dangerousness," 217–18. This high error rate is often noted in the literature. See, for example, Silver and Miller, "Cautionary Note," 142; von Hirsch and Gottfredson, "Selective Incapacitation"; Yang et al., "Efficacy." We return to the problem later.
46. Floud, "Dangerousness," 218–19.
47. Appelbaum, "New Preventive Detention."
48. Appelbaum, "New Preventive Detention."
49. It is important to note that criminal justice systems remain dependent upon actuarial assessment, which is increasingly recognized as quite distinct from algorithmic assessments premised upon big data. According to Hannah-Moffat, actuarial can be differentiated from algorithmic assessment according to epistemological and methodological dimensions. Epistemologically speaking, actuarial assessment predicts recidivism on the basis of sample populations, whereas algorithms, based on massive populations, can be marshaled to predict a wider range of outcomes. Methodologically, actuarial predictions depend upon specific types of data collection and variables; algorithms rely on data from multiple sources and search for relationships among a wide range of variables (see Hannah-Moffat, "Algorithmic Risk Governance," 458). Because criminal justice systems are not yet in the grip of algorithmic/big-data assessment, we prioritize actuarialism.
50. Mathiesen, "Selective Incapacitation Revisited," 465.
51. Cunningham and Reidy "Don't Confuse Me"; Dershowitz, "Preventive Detention"; Dvoskin and Heilbrun, "Risk Assessment"; Forst, "Selective Incapacitation"; Meehl, *Clinical versus Statistical*.
52. Shannon, "Risk Assessment," 161; Ewing, "Preventive Detention," 146; von Hirsch, "Prediction," 733–40.
53. On this point, see also the discussion in van Eijk, "Socioeconomic Marginality," 474–75.
54. See Forst, "Selective Incapacitation," 157–58; Kleiman et al., "Using Risk Assessment," 126; Skeem, "Risk Technology," 300. For a critique of construing "greater accuracy" as "ethical," see Mathiesen, "Selective Incapacitation Revisited," 465; Bueno, "Face Revisited," 75.

55. Gottfredson and Gottfredson, "Selective Incapacitation?"; Silver and Miller, "Cautionary Note," 141–43; von Hirsch, "Prediction of Criminal Conduct"; Wilkins, "Politics of Prediction."

56. Gottfredson and Gottfredson, "Selective Incapacitation," 145–47.

57. See also Blackmore and Welsh, "Selective Incapacitation," 516, in which it is noted that "high false positive rates have consistently bedevilled every effort to predict recidivism. False positive rates of about 50 to 60 percent have been the rule."

58. Gottfredson and Gottfredson, "Selective Incapacitation," 148.

59. Gottfredson and Gottfredson, "Selective Incapacitation," 149. Concerns with accuracy also underpin the critiques found in Gottfredson and Hirschi, "True Value"; Hirschi and Gottfredson, "Age," in which it is argued that inaccuracy in prediction stems from the inability to identify the "habitual criminal" prior to the production of a criminal record that verifies "habitual criminality." A similar point can also be found in Barnett and Lofaso, "Selective Incapacitation," 29.

60. A similar sentiment recurs in Berk, "Impact Assessment."

61. Berk, "Balancing the Costs," 1078.

62. Blackstone, *Commentaries on the Laws of England*, 2:358.

63. Berk, "Balancing the Costs," 1081 (emphasis in original).

64. On this point, see also Clear and Barry, "Some Conceptual Issues," 535. Clear and Barry posit that arriving at a model is easy enough, but difficulty arises insofar as there is no mechanism for "confirming the relative values placed on these types of errors."

65. See chapter 2. See also Mathiesen, "Selective Incapacitation Revisited," 467.

66. Oleson et al., "Training to See Risk," 55.

67. Kleiman et al, "Using Risk Assessment," 125–26; Monahan and Skeem, "Risk Assessment," 505–6. See also Hyatt et al., "Follow the Evidence."

68. Wolfgang et al., *Delinquency*; Greenwood and Abrahamse, *Selective Incapacitation*. See also Cohen, "Selective Incapacitation"; Mednick and Christiansen, *Biosocial Bases*; Wilson and Herrnstein, *Crime*.

69. Harcourt, *Against Prediction*, 122–23.

70. Harcourt, *Against Prediction*, 123–24.

71. Marvel and Moody, "Lethal Effects" (2001).

72. Kovandzic et al., "Unintended Consequences" (2002).

73. Zimring, "Populism."

74. Kovandzic et al., "Unintended Consequences"; Marvel and Moody, "Lethal Effects." See also Taifa, "Three-Strikes," 724. It might be worth drawing attention to Zimring, "Populism," 249–50, and Shichor, "Three Strikes," 481–83, in which it is argued that "three-strikes" legislation can lead to grossly disproportionate and inconsistent sentencing outcomes across similarly situated defendants, which would seem to be the flipside of "incentivization."

75. Blackmore and Welsh, "Selective Incapacitation," 508.

76. Clear and Barry, "Some Conceptual Issues," 530.

77. Clear and Barry, 539.

78. For the notion that prediction cannot but essentialize individuals as "offenders," thereby rendering them "unworthy" of concern, see also von Hirsch, "Prediction," 758; Ewing, "Preventive Detention"; Foote, "Comments."

79. Of course, we are assuming a reading of constitutional rights prior to being refracted by actuarialism. See Feeley and Simon, "Actuarial Justice," in which, as noted previously, it is argued that actuarialism recrafts the meaning of long-standing legal concepts and ideas.

80. It might be worth drawing attention to feminist arguments, in which strip searches of incarcerated women are, correctly in our view, understood as sexual abuse. See for example, Davis, *Are Prisons Obsolete?*, 77–83.

81. See, for example, Chaiken and Chaiken, *Varieties*; Sampsell-Jones, "Preventive Detention"; Barnett, "Getting Even"; Hruska, "Preventive Detention"; von Hirsch, "Prediction"; Hyatt et al., "Follow the Evidence"; Monahan, "Jurisprudence of Risk"; Morris, "'Dangerousness.'" Of course, scholars offer different resolutions to the problem of "reconciling" prediction-based and retroactive interventions.

82. Robinson, "Punishing Dangerousness," 1443–44; Shannon, "Risk Assessment."

83. Cunningham cited in Blackmore and Welsh, "Selective Incapacitation," 513–14.

84. Blackmore and Welsh, "Selective Incapacitation," 514.

85. Ericson, "Ten Uncertainties," 348.

86. Costanza-Chock, "Design Justice," 2–3.

87. Costanza-Chock, "Design Justice," 4.

88. Magnet and Rodgers, "Stripping," 111.

89. Costanza-Chock, "Design Justice," 4; Magnet and Rodgers, "Stripping," 102. See also Mason and Magnet, "Surveillance Studies," 106.

90. See Lyon, "Surveillance," 23, in which it is noted that judgments hinge on codes, which constitute "the switches that place one person in, say, the Affluentials category, and the next in Big City Stress, one person as having health risks, the next as having good prospects."

91. Henne and Shah, "Unveiling White Logic," 114.

92. Henne and Shah, "Unveiling White Logic," 115.

93. Henne and Shah, "Unveiling White Logic," 116.

94. Van Cleve and Mayes, "Criminal Justice," 409.

95. Van Cleve and Mayes, "Criminal Justice," 410.

96. Harcourt, *Against Prediction*. See also van Eijk, "Socioeconomic Marginality."

97. Oleson, "Risk in Sentencing"; Mathiesen, "Selective Incapacitation Revisited," 465; Harcourt, *Against Prediction* (2007).

98. Harcourt, *Against Prediction*, 145–71. See also Mathiesen, "Selective Incapacitation"; van Eijk, "Socioeconomic Marginality."

99. Brayne, "Big Data." See also Lyon, "Surveillance," 15–20.

100. Brayne, "Big Data," 987.

101. Brayne, "Big Data," 987. See also Ahmed, "Aided by Palantir."

102. Beckett et al., "Drug Use"; Epp et al., *Pulled Over*.

103. Brayne, "Big Data," 998.

104. Netter, "Using Group Statistics," 706. See also Silver and Miller, "Cautionary Note," 152.

105. Marx, *Capital*, 81-96.

106. Marx, *Capital*; see also Bataille, *Accursed Share.*

107. Mills, *Sociological Imagination*, 15.

108. Marx, *Capital*, 82.

109. Doyle, "Trust, Citizenship," 2. For further evidence of the tendency to read into actuarial assessment the possibility of "helping individuals" (and thus a progressive politics), see Silver and Miller, "Cautionary Note," 151.

CHAPTER 4. SECURITY SCIENCE

1. See Felson and Eckert, *Crime and Everyday Life.*

2. Coleman, *Criminal Elite*; Pratt and Eriksson, *Contrasts in Punishment*; Tombs and Whyte, *Corporate Criminal*; Wilkinson and Pickett, *Spirit Level.*

3. Coleman, *Criminal Elite.*

4. Barak, *Crimes by the Capitalist State.*

5. Coleman and Sim, "'You'll Never Walk Alone'"; Lippert and Wilkinson, "Capturing Crime," 134.

6. Mills, *Power Elite*; Fussell, *Class.*

7. Blalock et al., "Impact of Post-9/11."

8. Cannon, *Official Negligence.*

9. After the defendants were acquitted on April 29, 1992, the Los Angeles riots raged. The four officers were later prosecuted in federal court for violating King's civil rights (18 USC §242). See United States v. Koon, 34 F.3d 1416 (9th Cir. 1994); Koon v. United States, 518 U.S. 81 (1996) (applying US sentencing guidelines to Koon sentencing).

10. Hattery and Smith, *Policing Black Bodies*, 40.

11. On this point, see Goold, "Public Area Surveillance."

12. Alden, *Closing of American Border.*

13. Blalock et al., "Impact of Post-9/11"; Poole, "Airport Security"; Seidenstat, "Terrorism, Airport Security."

14. "Shoe Bomb Suspect."

15. Indictment, *United States v. Abdulmutallab* (2010).

16. Jansen, "TSA Defends."

17. Bunker and Flaherty, *Body Cavity Bombers*; *The Telegraph*, "Terrorist Hid Explosives."

18. Dearden, "Isis"; Muir and Cowan, "Four Bombs" (airports and subway stations); Samti and Gall, "Scores Die in Attack"; BBC News, "Bali Death Toll" (holiday resorts); *The Guardian*, "London Terror Attack"; Jehl, "70 Die"; and Michaelson, "Roadside Bomb" (tourist attractions); BBC News, "What Happened" (public events); and Withnall and Lichfield, "Charlie Hebdo" (newspapers).

19. Roussell, "Policing the Anticommunity," 820.

20. Roussell, "Policing the Anticommunity," 821.

21. Hulsman, "Critical Criminology."

22. Mainstream criminology is also interested in hierarchical taxonomies. See, for example, Sellin and Wolfgang, *Measurement of Delinquency* and a line of contemporary work in this tradition.

23. Wray, *Preliminary Semiannual.*

24. Norris et al., "Growth of CCTV," 111; Sivarajasingam et al., "Effect of Urban"; Wilson and Sutton, "Watched Over."

25. Joh, "Forgotten Threat," 386; Shearing and Stenning, "Modern Private Security," 215.

26. We return to broken windows and order maintenance policing in the next chapter, which focuses on environmental crime science. It is worth noting that these latter policing strategies may also appear to hinge upon dissipating and reterritorializing criminalized behaviors, but they are dependent upon different conceptual techniques for extending control over relatively innocuous behavior. Broken windows, for instance, portrays "minor crime" and "disorder" as an invitation to "serious crime." In this, vertical, hierarchical taxonomies of crime need not be disrupted. Rather, a line is drawn that links the "minor" to the "serious," often leading to a concentration of resources upon the former. See Wilson and Kelling, "Broken Windows."

27. Martin, "Bombs, Bodies," 17. This is, of course, the realization of Ben Stiller's "You can't say 'bomb' on an airplane . . . bomb bomb bomb, bomb bomb ba-bomb" scene from the pre-9/11 film *Meet the Parents.*

28. Hobbes, *Leviathan.*

29. United States v. Carroll Towing Co., 159 F.2d 169 (2d. Cir. 1947).

30. Landes, "Economic Study."

31. Landes, "Economic Study," 29.

32. Landes, "Economic Study," 4 (Cuba); and National Commission on Terrorist Attacks upon the United States, *9/11 Commission Report* (collapse skyscrapers).

33. Klein, *Shock Doctrine*, 7.

34. Loader, "Thinking Normatively."

35. For example, van Steden and Sarre, "Growth of Private Security."

36. Coleman and Sim, "You'll Never Walk Alone.'"

37. Congress passed the Authorization for the Use of Military Force against Terrorists (Pub. L. 107–40, 115 Stat. 224), thereby authorizing "the President to use all necessary and appropriate force against those nations, organizations, or persons he determines planned, authorized, committed, or aided the terrorist attacks that occurred on September 11, 2001, or harbored such organizations or persons, in order to prevent any future acts of international terrorism against the United States." The AUMF has been interpreted as a blank check to make war. It does not expire, and by early 2019, it had been invoked at least forty-one times in nineteen different countries.

38. Klein, *Shock Doctrine*, 10.

39. Wikipedia, s.v. "G4S."

40. Statista, "Revenue of G4S."

41. Risen, *Pay Any Price*, 33.

42. Risen, *Pay Any Price*, 48.

43. Risen, *Pay Any Price*, 35, 40.

44. Risen, *Pay Any Price*, 37–46.

45. Schneier, *Beyond Fear*, 38.

46. Gill and Turbin, "CCTV and Shop Theft"; Hancox and Morgan, "Use of CCTV"; Poyner, "Situational Crime." See also Clarke, *Situational Crime.*

47. Gill and Spriggs, *Assessing the Impact*; Haggerty, "Surveillance."

48. Norris, "Success of Failure"; Sivarajasingam et al., "Effect of Urban," 314. See also Bergen et al., "Do NSA's Bulk Surveillance" for the argument that NSA surveillance also has no deterrent effect.

49. Leman-Langlois, "Local Impact," 44.

50. Fisher, "Surveillance Auckland."

51. Carlile, "In Britain, Somebody's Watching You."

52. For the figures provided, see Norris et al., "Growth of CCTV"; Norris, "Success of Failure," 252.

53. Davis, "Beyond Blade Runner."

54. Davis, "Beyond Blade Runner,"n.p.

55. Davis, "Beyond Blade Runner," n.p. See also Davis, *City of Quartz.*

56. Haggerty, "Surveillance"; Leman-Langlois, "Local Impact," 38.

57. Engels, *Condition of the Working Classes.*

58. Harvey, "Fetish of Technology," 3.

59. Clarke, *Profiles of the Future*, 36.

60. Oleson, "Drown the World," 71.

61. Toffler, *Future Shock.*

62. Harvey, "Fetish of Technology," 23.

63. Harvey, "Fetish of Technology," 15.

64. On the PATRIOT Act, see Electronic Privacy Information Center, Uniting and Strengthening America . . . Act.

65. Blumenthal, "Congress Had No Time to Read."

66. ACLU, "Surveillance." Conversely, standard law enforcement procedures, rather than mass surveillance programs, have been the most effective strategy against terrorism. See Bergen et al., "Do NSA's Bulk Surveillance."

67. Greenwald, *No Place.*

68. Jordan, "Decrypting," 511–14.

69. Kadidal, "NSA Surveillance," 457.

70. Kadidal, "NSA Surveillance," 473.

71. Wikileaks, "Vault 7".

72. Solzhenitsyn, *Cancer Ward*, 208.

73. Orwell, *Nineteen Eighty-Four.*

74. Pedersen, "China Trials 'Deadbeat Map.'"

75. Orwell, *Nineteen Eighty-Four*, 267.

76. Harvey, "Fetish of Technology," 3–4.

77. Cf. Feeley and Simon, "New Penology."

78. For the fetish that surrounds CCTV, see Davies and Velastin, "Progress Review." For the fetish surrounding airport security, see Frederickson and LaPorte, "Airport Security"; Schwaninger, "Increasing Efficiency."

79. Landes, "Economic Study" (hijackings); and Cauley and Im, "Intervention Policy"; Enders and Sandler, "Effectiveness"; Enders and Sandler, "What Do We Know."

80. Cauley and Im, "Intervention Policy," 30.

81. Enders and Sandler, "Effectiveness," 843.

82. Cauley and Im, "Intervention Policy," 28.

83. "Shoe Bomb Suspect"; Indictment, United States v. Abdulmutallab (2010).

84. Dearden, "Isis"; Muir and Cowan, "Four Bombs"; *The Guardian*, "London Terror Attack"; Jehl, "70 Die"; Michaelson, "Roadside Bomb."

85. Fox News, "McVeigh's Apr. 26 Letter."

86. Parks, "Points of Departure" arrives at a similar conclusion in discussing the limits to contemporary airport security in the US.

87. Gras, "Legal Regulation"; Norris et al., "Growth of CCTV."

88. Lippert and Wilkinson, "Capturing Crime," 140.

89. Berndtsson and Stern, "Private Security"; Lippert and Walby, *Criminology of Policing*; Shearing and Stenning, "Modern Private Security," 238; van Steden and Sarre, "Growth of Private Security"; Wakefield, "Public Surveillance," 542–44.

90. Bloom and Dunn, "Constitutional Infirmity"; Collins, "And the Walls"; Evans, "Hijacking Civil Liberties"; Jaffer, "Balancing Power"; Kitrosser, "'Macro-Transparency'"; Whitehead and Aden, "Forfeiting."

91. Taylor, "State Surveillance"; Wilson and Sutton, "Watched Over" (CCTV); Collins, "And the Walls"; Conrad, "Executive Order"; Jordan, "Decrypting" (spy programs); Katyal and Caplan, "Surprisingly Stronger" (separation of powers); and Balkin, "Constitution"; Loader, "Thinking Normatively"; Sarre, "Researching Private Policing" (security science).

92. Cf. Dostoyevsky, *Brothers Karamazov*. In Dostoevsky's extraordinary parable of the Grand Inquisitor (p. 129), the Inquisitor condemns the returned Christ for expecting people to choose freedom over security:

> In the end they will lay their freedom at our feet, and say to us, "Make us your slaves, but feed us." They will understand themselves, at last, that freedom and bread enough for all are inconceivable together, for never, never will they be able to share between them! They will be convinced, too, that they can never be free, for they are weak, vicious, worthless, and rebellious. Thou didst promise them the bread of Heaven, but, I repeat again, can it compare with earthly bread in the eyes of the weak, ever sinful and ignoble race of man? And if for the sake of the bread of Heaven thousands shall follow Thee, what is to become of the millions and tens of thousands of millions of creatures who will not have the strength to forego the earthly bread for the sake of the heavenly?

93. Foucault, *Discipline and Punish*, 222.

94. Foucault, *Discipline and Punish*, 223 (emphasis added).

95. Similar arguments have been advanced by Abrahamsen and Williams, "Security Sector," and Abrahamsen and Williams, "Securing the City." They point out that state policing mechanisms and security technologies can often operate as a "hybrid" force in the maintenance of social order. However, they do not necessarily frame hybridization as evidence of "counter-legality" in the sense we adopt here.

96. Foucault, *Discipline and Punish*, 213–15.

97. Concurring in the 1991 case California v. Acevedo (500 U.S. 565, 582), Supreme Court Justice Antonin Scalia claimed that the warrant requirement "had become so riddled with exceptions that it was basically unrecognizable."

98. Landynski, *Search and Seizure*, 19–48 ("writs of assistance"); and Katz v. United States, 389 U.S. 347, 357 (1967) ("well-delineated exceptions").

99. Acevedo, at 582.

100. Mapp v. Ohio, 367 U.S. 643 (1961) (exclusionary rule). See Sklansky, "Is the Exclusionary Rule Obsolete?" (exceptions to *that* rule).

101. Slobogin, "Testilying."
102. Cannon, "One Bad Cop."
103. Maas, *Serpico.*
104. Prenzler, *Police Corruption.*
105. Leo, "False Confessions."
106. Skolnick and Fyfe, *Above the Law.*
107. Burke, "Consent Searches."
108. Taibbi, *Divide.*
109. Graham, "Overcharging."
110. Stuntz, *Collapse of American Criminal Justice.*
111. National Association of Criminal Defense Lawyers, *Trial Penalty.*
112. 18 U.S.C. § 2331 et seq.
113. Feldman, "'Intelligence' and 'Evidence.'"
114. Oleson, "Drown the World," 61–70.
115. Rasdan, "Moussaoui Case," 1420.
116. United States v. Reynolds, 345 U.S. 1, 12 (1953).
117. Carter, "Court Order Violations."
118. Dwyer, "One Juror."
119. United States v. Moussaoui, No. CR 01-455-A, 2003 WL 21263699, at *6 (E.D. Va. filed Dec. 11, 2001).
120. See Taub, "Guantánamo's Darkest Secret."
121. Wallach, "Drop by Drop"; Form, "Charging Waterboarding" (IMTFE). See Greenberg and Dratel, *Torture Papers.*
122. Honigsberg, *Place Outside the Law* ("place outside the law"); Denbeaux and Hafetz, *Guantánamo Lawyers* ("a prison outside the law"); and Steyn, "Guantánamo Bay."
123. Bravin, *Terror Courts.*
124. See generally Rasul v. Bush, 542 U.S. 466 (2004) (US courts have jurisdiction to hear habeas corpus challenges from noncitizens detained at Guantánamo); Hamdi v. Rumsfeld, 542 U.S. 507 (2004) (US citizens are entitled to contest the designation of enemy combatant before a neutral decisionmaker).
125. Foley, "Guantanamo and Beyond," 1009–10.
126. Hussain, "Guantánamo Bay 'Forever' Prisoner."
127. Hamdan v. Rumsfeld, 548 U.S. 557 (2006).
128. Boumediene v. Bush, 553 U.S. 723 (2008).
129. Rosenberg, "Trial for Men."
130. Foley, "Guantanamo and Beyond."
131. Somaiya, "Tracking the News."
132. Shane, "U.S. Approves Targeted Killing."
133. Agamben, *Homo Sacer.*
134. For discussion of the state secret doctrine, see Fisher, *In the Name of National Security.*
135. Greenwald, "Obama Killed."
136. Boumediene, 553 U.S. 723, 797.
137. Schmitt, *Political Theology*, 5.
138. Agamben, *State of Exception*, 2.
139. Masferrer, *Post 9/11*; Ralph, *America's War on Terror*. But an ontologically more terrifying possibility is that the concentration camp is not an

exception to the modern social order, but a distillation of it. See Mrázek, *Complete Lives*.

140. Orwell, *Nineteen Eighty-Four*, 183.

141. Masco, "Secrecy/Threat Matrix," 183.

CHAPTER 5. ENVIRONMENTAL CRIME SCIENCE

1. Cohen and Felson, "Social Change," 598–600.

2. Clarke, *Situational Crime*, 6.

3. Cozens et al., "Crime Prevention"; Jeffery, *Crime Prevention*.

4. Kelling and Coles, *Fixing Broken Windows*; Wilson and Kelling, "Broken Windows."

5. Sherman and Weisburd, "General Deterrent Effects."

6. Silverman, *NYPD Battles*, 98.

7. Harvey, *Brief History*, 64–67, 78, 160–61.

8. Bichler and Clarke, "Eliminating Pay Phone Toll Fraud," 109.

9. Hunter and Jeffery, "Preventing Convenience Store Robbery," 199.

10. Hunter and Jeffery, "Preventing Convenience Store Robbery," 199.

11. Willis et al., "Making Sense of COMPSTAT," 165.

12. Eterno and Silverman, *Crime Numbers Game*.

13. Moore, "Sizing up Compstat," 492.

14. Smith, *Inquiry*.

15. Harvey, *Brief History*, 64–65; Kramer, "Moral Panics"; Logan and Molotch, *Urban Fortunes*; Smith, *New Urban Frontier*.

16. For these arguments, see Cozens et al., "Crime Prevention," 337; Ferrell, "Urban Graffiti"; Ferrell, "Freight Train Graffiti"; Jacobs et al., "Carjacking, Streetlife"; Jacobs, "Manipulation of Fear," 526; Wing, "Putting the Brakes On," 386. See also Massey et al., "Property Crime"; Moore, "Sizing up Compstat," 489; Palmen, "Forget Carjacking"; Tilley, "Evaluating the Effectiveness"; Willis et al., "Making Sense of Compstat," 174–75; Wilson and Sutton, "Watched Over," 221. It is also worth drawing attention to Reppetto, "Crime Prevention," who is often read as endorsing displacement, although he ultimately argues that "mechanical prevention" may be effective given that certain crimes are "so opportunistic in nature" (177).

17. Hsu and McDowall, "Does Target-Hardening Result."

18. Hsu and McDowall, "Does Target-Hardening Result," 944.

19. Hsu and McDowall, "Does Target-Hardening Result," 945–46.

20. Knutsson and Kuhlhorn, "Macro-Measures against Crime," 116.

21. Clarke, *Situational Crime*, 28.

22. Kuhlhorn, "Housing Allowances," 236, 240.

23. Wacquant, "Crafting the Neoliberal State."

24. Clarke, *Situational Crime*, 3.

25. Miethe and Sousa, "Carjacking," 257.

26. Miethe and Sousa, "Carjacking," 257.

27. Miethe and Sousa, "Carjacking," 244, 249, 254–55.

28. Miethe and Sousa, "Carjacking," 256.

29. Finkelhor and Asdigian, "Risk Factors for Youth," 6.

30. Miller et al., *Victim Costs.*

31. Finkelhor and Asdigian, "Risk Factors for Youth," 4–7.

32. Stanko, "Warnings to Women," 19.

33. Stanko, "Warnings to Women," 17 (emphasis in original).

34. Walklate, "Risk and Criminal Victimization," 38.

35. Walklate, "Risk and Criminal Victimization," 44.

36. The court's opinion in a 1981 Maryland case similarly constructs "natural instincts" as a mechanism to shift responsibility from the powerful to the powerless:

> While courts no longer require a female to resist to the utmost or to resist where resistance would be foolhardy, they do require her acquiescence in the act of intercourse to stem from fear generated by something of substance. She may not simply say "I was really scared" and thereby transform consent or mere unwillingness into submission by force. These words do not transform a seducer into a rapist. *She must follow the natural instinct of every proud female to resist,* by more than mere words, the violation of her person.

State v. Rusk, 424 A.2d 720 (1981) (emphasis added).

37. Wilson and Kelling, "Broken Windows," 4.

38. Wilson and Kelling, "Broken Windows," 5.

39. Friedman and Percival, *Roots of Justice*; Miller, *Cops and Bobbies*; Walker, "Broken Windows."

40. Walker, "Broken Windows," 86.

41. Reiss, *Police and the Public* (1971); Walker, "Broken Windows," 85.

42. Donner, *Protectors of Privilege* (1990).

43. Roussell, "Policing the Anticommunity.". See also Gascón and Roussell, *Limits of Community Policing*; Roussell and Gascón, "Defining 'Policeability.'"

44. Wilson and Kelling "Broken Windows," 2–3.

45. Wilson and Kelling "Broken Windows," 7. See also page 2, where it is noted that "some of the things [Officer Kelly] did probably would not withstand a legal challenge."

46. Wilson and Kelling "Broken Windows," 6.

47. See also Kelling, "Crime Control," in which it is argued that "abuses and corruption are inevitable in policing" and, as such, the best one can hope for is that such abuses be kept "to a minimum." The quoted material appears on page 8.

48. Bernasco and Luykx, "Effects of Attractiveness," 988 ("ethnic heterogeneity"); Roncek and Maier, "Bars, Blocks," 734 ("socioeconomic status"); and Tseloni et al., "Burglary Victimization," 83 ("household annual income").

49. Henne and Shah, "Unveiling White Logic"; Van Cleves and Mayes, "Criminal Justice."

50. Bernasco and Luykx, "Effects of Attractiveness," 987.

51. Bernasco and Luykx, "Effects of Attractiveness," 986.

52. Bernasco and Luykx, "Effects of Attractiveness," 995.

53. Bernasco and Luykx, "Effects of Attractiveness," 996.

54. Tseloni et al., "Burglary Victimization," 87.

55. Bernasco and Luykx, "Effects of Attractiveness," 84.

56. Bernasco and Luykx, "Effects of Attractiveness," 84.

57. Fagan and Davies, "Street Stops." We use this article for illustrative purposes, but by no means is it an isolated example. Even in accounts that are critical in many respects, one can find recuperations of the broken windows thesis. See Collins, "Strolling While Poor," 438; Fagan et al., "Street Stops" (2010); Walker, "Broken Windows," 88.

58. Fagan and Davies, "Street Stops," 463.

59. Fagan and Davies, "Street Stops," 477.

60. Fagan and Davies, "Street Stops," 478.

61. Fagan and Davies, "Street Stops," 482. See also page 489.

62. Fagan and Davies, "Street Stops," 464, 495.

63. Fagan and Davies, "Street Stops," 469–72.

64. Fagan and Davies, "Street Stops," 496.

65. Wilson and Kelling, "Broken Windows," 1.

66. Wilson and Kelling, "Broken Windows," 3.

67. Wilson and Kelling, "Broken Windows," 6. See also Kelling and Coles, *Fixing Broken Windows*.

68. Fagan and Davies, "Street Stops."

69. Wilson and Kelling, "Broken Windows," 6.

70. Cf. Feeley and Simon, "New Penology."

71. Wilson and Kelling, "Broken Windows," 6.

72. Wilson and Kelling, "Broken Windows," 6.

73. Concerning fractured communities, it is worth recalling the work of Gascón and Roussell, *Limits of Community Policing*; Roussell, "Policing the Anticommunity"; Roussell and Gascón, "Defining 'Policeability.'"

74. Eterno and Silverman, "New York City," 219; Weisburd et al., "Growth of Compstat"; Weisburd et al., "Reforming to Preserve," 433; Willis et al., "Compstat and Bureaucracy"; Willis et al., "Making Sense," 161.

75. See, for example, Eterno and Silverman, "New York City"; Howell, "Broken Lives"; Moore, "Sizing up Compstat"; Walsh, "Compstat."

76. Howell "Broken Lives," 281–83.

77. Howell "Broken Lives," 282.

78. Howell "Broken Lives," 282.See also Harcourt and Ludwig, "Reefer Madness."

79. Howell, "Broken Lives," 291.

80. Howell, "Broken Lives," 293–305. See also Alexander, *New Jim Crow*; Jacobs, *Eternal Criminal Record*.

81. Eterno and Silverman, "New York City," 225.

82. For these figures, see Eterno and Silverman, "New York City," 222; Walsh, "Compstat," 359.

83. Tyler, *Why People Obey the Law*.

84. Howell provides a fairly extensive list of criminal justice reforms that, if implemented, could mitigate the damaging effects that broken windows policing have on individuals. See Howell, "Broken Lives," 315–25.

85. Eterno and Silverman, "New York City," 221.

86. Eterno and Silverman, "New York City," 227. For a similar sentiment, see also Moore, "Sizing up Compstat," 492.

87. Cohen and Felson, "Social Change," 589.

88. Bernasco and Luykx, "Effects of Attractiveness," 985.

89. Clarke, "Policing Terrorism," 19.

90. Clarke, "Policing Terrorism," 19.

91. Clarke, "Policing Terrorism," 22 ("political controversies surrounding the barrier") and 7 ("waste time on motives"). See also Perry et al., "Situational Prevention."

92. Our emphasis on "adaptation" would be quite consistent with some of those who promote CPTED, such as Jeffery, *Crime Prevention*. The meanings and implications that it has for us, however, are distinct.

93. Clarke, *Situational Crime*. The quoted material is from pages 38 and 4, respectively.

94. See Clarke, *Situational Crime*. The text demonstrates these points.

95. Clarke, *Situational*, 40. See also O'Malley, *Crime and Risk*, who buys into the idea that preemptive strategies ought to be appreciated for their potential to be nonpunitive.

96. For a somewhat similar interpretation, see White, "Graffiti," 265.

97. Cohen and Felson, "Social Change," 605.

98. A similar problem plagues "biosocial criminology." See chapter 2.

99. Clarke, *Situational Crime*, 5.

100. Clarke, *Situational Crime*, 42.

101. Clarke, *Situational Crime*, 42–43.

102. Masuda, "Reduction of Employee Theft."

103. DiLonardo, "Economic Benefit," 124.

104. Concerning the former, it is worth recalling the work of Finkelhor and Asdigian, "Risk Factors for Youth"; Stanko, "Warnings to Women"; and Walklate, "Risk and Criminal Victimization," discussed in an earlier section of this chapter. Concerning the latter, see Walters, "Bhopal, Corporate Crime"; Bradshaw, "Blacking Out the Gulf."

105. Clarke, *Situational Crime*, 41.

106. Clarke, *Situational Crime*, 41.

107. Clarke, *Situational Crime*, 41.

108. Geason, "Preventing Graffiti," 2.

109. Foucault, *Discipline and Punish*.

CONCLUSION

1. Lombroso, *Criminal Man*, 171. See also chapter 1.

2. Carrier, "Academics' Criminals"; Carrier, "Speech for the Defense."

3. Gerlach, *Genetic Imaginary*.

4. Lombroso, *Criminal Man*, 92.

5. Lombroso, *Criminal Man*, 268.

6. Walsh, *Race and Crime*; Walsh and Beaver, "Introduction."

7. Hannah-Moffat, "Criminogenic Needs"; Harcourt, *Against Prediction*; van Eijk, "Socioeconomic Marginality" (proxies for race and class). See Cole, *Suspect Identities*, 258 ("immaculate").

8. Wilson and Kelling, "Broken Windows."

9. See Gerlach, *Genetic Imaginary*. See also chapter 2.

10. Martin, "Bombs, Bodies."

11. Wilson and Kelling, "Broken Windows"; Clarke, *Situational Crime*.

12. Walsh, *Race and Crime*; Walsh and Beaver, "Introduction," 14. See also Rowe, *Biology and Crime.*

13. Henne and Shah, "Unveiling White Logic"; Van Cleve and Mayes, "Criminal Justice."

14. These tendencies are especially evident in Clarke, *Situational Crime.*

15. Beccaria, *On Crimes*, 15; Lombroso, *Criminal Man*, 268.

16. See, for example, Feeley and Simon, "Actuarial Justice"; Lynch et al., *Truth Machine.*

17. Gerlach, *Genetic Imaginary*, 91, 134; Kruse, *Social Life*, 124–30; Lynch et al., *Truth Machine*, 235. See also Cole, *Suspect Identities*, 63–65, 96.

18. Floud, "Dangerousness," 218–19.

19. Agamben, *Homo Sacer*; Agamben, *State of Exception*, 2; Masferrer, *Post 9/11*; Mrázek, *Complete Lives*; Ralph, *America's War on Terror.*

20. Kelling, "Crime Control."

21. Walsh and Beaver "Introduction," 7.

22. On biosocial criminology's recapitulation of (Lombrosian) biological criminology, see especially Carrier and Walby, "Ptolemizing Lombroso." See also Walby and Carrier, "Rise of Biocriminology."

23. Clarke, *Situational Crime*, 41. See also Geason, "Preventing Graffiti," 2.

24. Risen, *Pay Any Price.*

25. Norris et al., "Growth of CCTV."

26. Cullen et al., "Prisons Do Not Reduce Recidivism."

27. Durose et al., *Recidivism of Prisoners.*

28. Wagner and Rabuy, "Following the Money."

29. Oleson, "Habitual Criminal Legislation."

30. See Carrier, "Academics' Criminals."

31. Bourdieu, *Outline of a Theory.*

32. Walsh and Wright, "Rage against Reason."

33. Russell, "Failure of Postmodern Criminology." See also Henry and Milovanovic, *Constitutive Criminology.*

34. Walsh, *Race and Crime.*

35. Haraway, "Situated Knowledges."

36. Carrier, "Speech for the Defense," 173.

37. Walsh, *Race and Crime.* See also Walsh and Wright, "Rage against Reason," 64; chapter 2.

38. Kruse, *Social Life*, 135.

39. See especially Berk, "Balancing the Costs."

40. Alexander, *New Jim Crow*; Wacquant, *Prisons of Poverty* (2009); Western, *Punishment and Inequality.*

41. Since 2011, the annual reports of the Global Commission on Drug Policy have stated unequivocally that the war on drugs has failed. Global Commission on Drug Policy, *War on Drugs.* In subsequent reports, the GCDP has assessed the international evidence and called for decriminalization (2016) and responsible regulation of drug supply (2018).

42. Carrier, "Speech for the Defense."

43. Henne and Shah, "Unveiling White Logic"; Van Cleve and Mayes, "Criminal Justice."

Bibliography

Abrahamsen, R., and Williams, M. C. "Securing the City: Private Security Companies and Non-state Authority in Global Governance." *International Relations* 21, no. 2 (2007): 237–53.

———. "Security Sector Reform: Bringing the Private In." *Conflict, Security and Development* 6, no. 1 (2006): 1–23.

Agamben, G. *Homo Sacer: Sovereign Power and Bare Life*. Translated by D. Heller-Roazen. Stanford, CA: Stanford University Press, 1998.

———. *State of Exception*. Translated by K. Attell. Chicago: University of Chicago Press, 2005.

Ahmed, M. "Aided by Palantir, the LAPD Uses Predictive Policing to Monitor Specific People and Neighborhoods." The Intercept, May 11, 2018. Accessed February 7, 2019. https://theintercept.com/2018/05/11/predictive-policing-surveillance-los-angeles/.

Alden, E. H. *The Closing of the American Border: Terrorism, Immigration, and Security since 9/11*. New York: Harper, 2008.

Alexander, M. *The New Jim Crow: Mass Incarceration in the Age of Colorblindness*. New York: The New Press, 2012.

American Civil Liberties Union (ACLU). "Surveillance under the PATRIOT Act." Accessed June 24, 2020. www.aclu.org/issues/national-security/privacy-and-surveillance/surveillance-under-patriot-act.

Andrews, D. A., Bonta, J., and Wormith, S. J. *The Level of Service/Case Management Inventory (LS/CMI)*. Toronto: Multi-Health Systems, 2004.

Appelbaum, P. S. "The New Preventive Detention: Psychiatry's Problematic Responsibility for the Control of Violence." *American Journal of Psychiatry* 145, no. 7 (1988): 779–85.

Aronson, J. D., and Cole, S. A. "Science and the Death Penalty: DNA, Innocence, and the Debate over Capital Punishment in the United States." *Law and Social Inquiry* 34, no. 3 (2009): 603–33.

Baker, L. A., Jacobson, K. C., Raine, A., Lozano, D. I., and Bezdjian, S. "Genetic and Environmental Bases of Childhood Antisocial Behavior: A Multi-informant Twin Study." *Journal of Abnormal Psychology* 116, no. 2 (2007): 219–35.

Balkin, J. M. "The Constitution in the National Surveillance State." *Minnesota Law Review* 93 (2008): 1–25.

Barak, G. *Crimes by the Capitalist State: An Introduction to State Criminality*. Albany: State University of New York Press, 1991.

Barlow, H. D. *Introduction to Criminology*. 6th ed. New York: HarperCollins, 1993.

Barnett, A., and Lofaso, A. J. "Selective Incapacitation and the Philadelphia Cohort Data." *Journal of Quantitative Criminology* 1, no. 1 (1985): 3–36.

Barnett, R. E. "Getting Even: Restitution, Preventive Detention, and the Tort/Crime Distinction." *Boston University Law Review* 76 (1996): 157–68.

Bataille, G. *The Accursed Share*. Vol. 1. Translated by R. Hurley. New York: Zone Books, 1991.

BBC News. "Bali Death Toll Set at 202." February 19, 2003. Accessed June 24, 2020. http://news.bbc.co.uk/2/hi/asia-pacific/2778923.stm.

———. "What Happened at the Bataclan?" December 9, 2015. Accessed June 24, 2020. www.bbc.com/news/world-europe-34827497.

Beccaria, C. *On Crimes and Punishments*. Indianapolis, IN: Bobbs-Merrill, 1963.

Becker, P., and Wetzell, R. F. *Criminals and Their Scientists: The History of Criminology in International Perspective*. New York: Cambridge University Press, 2006.

Beckett, K., Nyrop, K., Pfingst, L., and Bowen, M. "Drug Use, Drug Possession Arrests, and the Question of Race: Lessons from Seattle." *Social Problems* 52, no. 3 (2005): 419–41.

Bentham, J. *An Introduction to the Principles of Morals and Legislation*. Oxford: Clarendon Press, 1879.

Berg, S. O. "Sherlock Holmes: Father of Scientific Crime Detection." *Journal of Criminal Law, Criminology and Police Science* 61, no. 3 (1970): 446–52.

Bergen, P., Sterman, D., Schneider, E., and Cahall, B. "Do NSA's Bulk Surveillance Programs Stop Terrorists?" *New America Foundation* (January 2014): 1–32.

Berger, P. L., and Luckmann, T. *The Social Construction of Reality: A Treatise in the Sociology of Knowledge*. Garden City, NY: Anchor Books, 1966.

Bergland, Christopher. "One Simple Way You Can Become a Human Lie Detector." *The Athlete's Way* (blog). *Psychology Today*, October 31, 2015. www.psychologytoday.com/nz/blog/the-athletes-way/201510/one-simple-way-you-can-become-human-lie-detector.

Berk, R. "Balancing the Costs of Forecasting Errors in Parole Decisions." *Albany Law Review* 74, no. 3 (2010/2011): 1071–85.

———. "An Impact Assessment of Machine Learning Risk Forecasts on Parole Board Decisions and Recidivism." *Journal of Experimental Criminology* 13, no. 2 (2017): 193–216.

Berman, G., and Feinblatt, J. *Good Courts: The Case for Problem-Solving Justice*. New York: The New Press, 2005.

Bernasco, W., and Luykx, F. "Effects of Attractiveness, Opportunity and Accessibility to Burglars on Residential Burglary Rates of Urban Neighborhoods." *Criminology* 41, no. 3 (2003): 981–1002.

Berndtsson, J., and Stern, M. "Private Security and the Public-Private Divide: Contested Lines of Distinction and Modes of Governance in the Stockholm-Arlanda Security Assemblage." *International Political Sociology* 5 (2011): 408–25.

"Beyond Blade Runner: Urban Control (2): The Ecology of Fear." *Mediamatic* 8, nos. 2/3, January 1, 1995. www.mediamatic.net/en/page/8924/beyond-blade-runner-urban-control-2.

Bichler, G., and Clarke, R. V. "Eliminating Pay Phone Toll Fraud at the Port Authority Bus Terminal in Manhattan." In *Situational Crime Prevention: Successful Case Studies*, edited by R. V. Clarke, 98–112. Monsey, NY: Criminal Justice Press, 1997.

Blackmore, J., and Welsh, J. "Selective Incapacitation: Sentencing According to Risk." *Crime and Delinquency* 29 (1983): 504–28.

Blackstone, W. *Commentaries on the Laws of England in Four Books: Notes Selected from the Editions of Archibold, Christian, Coleridge, Chitty, Stewart, Kerr, and Others, Barron Field's Analysis, and Additional Notes, and a Life of the Author by George Sharswood; In Two Volumes*. Philadelphia: J. B. Lippincott, 1893.

Blalock, G., Kadiyali, V., and Simon, D. H. "The Impact of Post-9/11 Airport Security Measures on the Demand for Air Travel." *Journal of Law and Economics* 50, no. 4 (2007): 731–55.

Bloom, R. M., and Dunn, W. J. "The Constitutional Infirmity of Warrantless NSA Surveillance: The Abuse of Presidential Power and the Injury to the Fourth Amendment." *William and Mary Bill of Rights Journal* 15 (2007): 147–202.

Blumenthal, P. "Congress Had No Time to Read the USA PATRIOT Act." Sunlight Foundation, March 2, 2009. Accessed June 24, 2020. https://sunlightfoundation.com/2009/03/02/congress-had-no-time-to-read-the-usa-patriot-act/

Bonta, J., and Andrews, D. A. "Risk-Need-Responsivity Model for Offender Assessment and Rehabilitation." Public Safety Canada, 2007. Accessed February 5, 2019. www.publicsafety.gc.ca/cnt/rsrcs/pblctns/rsk-nd-rspnsvty/index-en.aspx.

Bourdieu, P. *Outline of a Theory of Practice*. Cambridge: Cambridge University Press, 1977.

Bradshaw, E. A. "Blacking Out the Gulf: State-Corporate Environmental Crime and the Response to the 2010 BP Oil Spill." In *The Routledge International Handbook of the Crimes of the Powerful*, edited by G. Barak, 383–92. New York: Routledge, 2015.

Bravin, J. *Terror Courts: Rough Justice at Guantánamo Bay*. New Haven, CT: Yale University Press, 2013.

Brayne, S. "Big Data Surveillance: The Case of Policing." *American Sociological Review* 82, no. 5 (2017): 977–1008.

Brunner, H. G., Nelen, M., Breakefield, X. O., Ropers, H. H., and van Oost, B. A. "Abnormal Behavior Associated with a Point Mutation in the Structural Gene for Monoamine Oxidase A." *Science* 262 (1993): 578–80.

Bueno, C. C. "The Face Revisited: Using Deleuze and Guattari to Explore the Politics of Algorithmic Face Recognition." *Theory, Culture and Society* 37, no. 1 (2020): 73–91.

Bunker, R. J., and Flaherty, C. *Body Cavity Bombers: The New Martyrs*. Bloomington, IN: iUniverse, 2013.

Burke, A. S. "Consent Searches and Fourth Amendment Reasonableness." *Florida Law Review* 67, no. 2 (2015): 509–63.

Cannon, L. "One Bad Cop." *New York Times Magazine*, October 1, 2000.

Cannon, L. *Official Negligence: How Rodney King and the Riots Changed Los Angeles and the LAPD*. Boulder, CO: Westview Press, 1999.

Carlile, J. "In Britain, Somebody's Watching You." NBC News, September 14, 2004. Accessed June 24, 2020. www.nbcnews.com/id/5942513/ns/world_news/t/britain-somebodys-watching-you/#.XLfkeS-Q2i4.

Carrier, N. "Academics' Criminals." *Champ Pénal/Penal Field* 3 (2006): 1–20.

———. "Anglo-Saxon Sociologies of the Punitive Turn: Critical Timidity, Reductive Perspectives, and the Problem of totalization." *Champ Pénal/Penal Field* 12 (2010): 1–19.

———. "Critical Criminology Meets radical constructivism." *Critical Criminology* 19, no. 4 (2011): 331–50.

———. "Speech for the Defense of a Radically Constructivist Sociology of (Criminal) Law." *International Journal of Law, Crime and Justice* 36, no. 3 (2008): 168–83.

Carrier, N., and Walby, K. "Ptolemizing Lombroso: The Pseudo-Revolution of Biosocial Criminology." *Journal of Theoretical and Philosophical Criminology* 1, no. 1 (2014): 1–45.

Carter, F. "Court Order Violations, Witness Coaching, and Obstructing Access to Witnesses: An Examination of the Unethical Attorney Conduct That Nearly Derailed the Moussaoui Trial." *Georgetown Journal of Legal Ethics* 20 (2007): 463–74.

Cartwright, J. *Evolution and Human Behavior: Darwinian Perspectives on Human Nature*. Cambridge, MA: MIT Press, 2000.

Cauley, J., and Im, E. I. "Intervention Policy Analysis of Skyjackings and Other Terrorist Incidents." *American Economic Review* 78, no. 2 (1988): 27–31.

Chaiken, J. M., and Chaiken, M. R. *Varieties of Criminal Behavior*. Santa Monica, CA: Rand Corporation, 1982.

Clarke, A. C. *Profiles of the Future: An Inquiry into the Limits of the Possible*. New York: Harper and Row, 1973.

Clarke, R. V. "Crime Science." In *The Sage Handbook of Criminological Theory*, edited by E. McLaughlin and T. Newbern, 271–83. Thousand Oaks, CA: Sage, 2010.

———. "Policing Terrorism: Principles of Target Hardening and Evaluated Examples." Paper presented at the 26th Problem-Oriented Policing Conference, Tempe, Arizona, October 24–26, 2016. Accessed February 15, 2019. https://popcenter.asu.edu/sites/default/files/policing_terrorism_clarke.pdf.

———. *Situational Crime Prevention: Successful Case Studies*. 2nd ed. Monsey, NY: Criminal Justice Press, 1997.
Clear, T. R., and Barry, D, M. "Some Conceptual Issues in Incapacitating Offenders." *Crime and Delinquency* 29 (1983): 529–45.
CNN. "Roberts: 'My Job Is to Call Balls and Strikes and Not to Pitch or Bat.'" September 12, 2005. Accessed June 24, 2020. http://edition.cnn.com/2005/POLITICS/09/12/roberts.statement/.
Cohen, J. "Selective Incapacitation: An Assessment." *University of Illinois Law Review* 2 (1984): 253–90.
Cohen, L. E., and Felson, M. "Social Change and Crime Rate Trends: A Routine Activity Approach." *American Sociological Review* 44 (1979): 588–608.
Cole, S. A. "More Than Zero: Accounting for Error in Latent Fingerprint Identification." *Journal of Criminal Law and Criminology* 95, no. 3 (2005): 985–1078.
———. "The 'Opinionization' of Fingerprint Evidence." *BioSocieties* 3 (2008): 105–13.
———. *Suspect Identities: A History of Fingerprinting and Criminal Identification*. Cambridge, MA: Harvard University Press, 2002.
Cole, S. A., and Dioso-Villa, R. "*CSI* and Its Effects: Media, Juries, and the Burden of Proof." *New England Law Review* 41 (2007): 435–70.
Cole, S. A., and Dioso-Villa, R. "Should Judges Worry about the 'CSI Effect?'" *Court Review* 47 (2011): 20–31.
Coleman, J. W. *The Criminal Elite: Understanding White-Collar Crime*. 6th ed. New York: Worth Publishers, 2005.
Colemen, R., and Sim, J. "'You'll Never Walk Alone': CCTV Surveillance, Order and Neoliberal Rule in Liverpool City Centre." *British Journal of Sociology* 51, no. 4 (2000): 623–39.
Collins, J. M. "And the Walls Came Tumbling Down: Sharing Grand Jury Information with the Intelligence Community under the USA Patriot Act." *American Criminal Law Review* 39 (2002): 1261–86.
Collins, R. "Strolling while Poor: How Broken-Windows Policing Created a New Crime in Baltimore." *Georgetown Journal on Poverty Law and Policy* 14, no. 3 (2007): 419–39.
Conrad, S. J. "Executive Order 12,333: Unleashing the CIA Violates the Leash Law." *Cornell Law Review* 70, no. 5 (1985): 968–90.
Costanza-Chock, S. "Design Justice, A.I., and Escape from the Matrix of Domination." MIT Open Access Articles, 2008. https://hdl.handle.net/1721.1/123083.
Cozens, P. M., Saville, G., and Hillier, D. "Crime Prevention through Environmental Design (CPTED): A Review and Modern Bibliography." *Property Management* 23, no. 5 (2005): 328–56.
Cullen, F. T., Jonson, C. L., and Nagin, D. S. "Prisons Do Not Reduce Recidivism: The High Cost of Ignoring Science." *Prison Journal* 91, no. 3 (supp. 2011): 48S–65S.
Cunneen, C. "Racism, Discrimination and the Over-Representation of Indigenous People in the Criminal Justice System: Some Conceptual and Explanatory Issues." *Current Issues in Criminal Justice* 17, no. 3 (2006): 329–46.

Cunningham, M., and Reidy, T. "Don't Confuse Me with the Facts: Common Errors in Violence Risk Assessment at Capital Sentencing." *Criminal Justice and Behavior* 26, no. 1 (1999): 20–43.

Darwin, C. *The Origin of Species by Means of Natural Selection*. New York: Barnes and Noble Classics, 2004.

Davies, A. C., and Velastin, S. A. "A Progress Review of Intelligent CCTV Surveillance Systems." *Proceedings IEEE IDAACS* (2005): 417–23.

Davis, A. *Are Prisons Obsolete?* New York: Seven Stories Press, 2003.

Davis, M. "Beyond Blade Runner: Urban Control (1)." *Mediamatic Magazine* 8, nos. 2/3, January 1, 1995. Accessed June 24, 2020. www.mediamatic.net/en/page/8923/beyond-blade-runner-urban-control-1.

———. *City of Quartz: Excavating the Future of Los Angeles*. London: Verso, 1990.

Dearden, L. "Isis Supporters Claim Group Responsible for Brussels Attacks: 'We Have Come to You with Slaughter.'" *Independent*, March 22, 2016. Accessed June 24, 2020. www.independent.co.uk/news/world/europe/isis-supporters-claim-responsibility-for-brussels-attacks-bombings-belgium-airport-maalbeek-metro-we-a6945886.html.

Deleuze, G. "Postscript on the Societies of Control." *October* 59 (1992): 3–7.

Deleuze, G., and Guattari, F. *A Thousand Plateaus: Capitalism and Schizophrenia*. Translated by B. Massumi. Minneapolis: University of Minnesota Press, 1987.

Denbeaux, M. P., and Hafetz, J. *The Guantánamo Lawyers: Inside a Prison outside the Law*. New York: New York University Press, 2009.

Dershowitz, A. "Preventive Detention and the Prediction of Dangerousness—The Law of Dangerousness: Some Fictions about Predictions." *Journal of Legal Education* 23 (1970): 24–47.

DiLonardo, R. L. "The Economic Benefit of Electronic Article Surveillance." In *Situational Crime Prevention: Successful Case Studies*, edited by R. V. Clarke, 122–31. Monsey, NY: Criminal Justice Press, 1997.

Donner, F. *Protectors of Privilege: Red Squads and Police Repression in Urban America*. Berkeley: University of California Press, 1990.

Dostoyevsky, F. *The Brothers Karamazov*. Translated by C. Garnett. Mineola, NY: Dover, 2005.

Douglas, M. *Purity and Danger: An Analysis of the Concepts of Pollution and Taboo*. London: Ark Paperbacks, 1984.

Doyle, A. "Trust, Citizenship and Exclusion in the Risk Society." *Law Commission of Canada, Risk and Trust: Including or Excluding Citizens* (2007): 7–22.

Doyle, A. C. "The Boscombe Valley Mystery." In *The New Annotated Sherlock Holmes*, Vol. 1, edited by L. S. Klinger, 101–32. New York: W. W. Norton, 2005.

Dror, I. E., and Mnookin, J. L. "The Use of Technology in Human Expert Domains: Challenges and Risks Arising from the Use of Automated Fingerprint Identification Systems in Forensic Science." *Law, Probability and Risk* 9 (2010): 47–67.

Durose, M. R., Cooper, A. D., and Snyder, H. M. *Recidivism of Prisoners Released in 30 States in 2005: Patterns from 2005 to 2010*. Washington, DC: Bureau of Justice Statistics, 2014.

Duster, T. *Backdoor to Eugenics*. New York: Routledge, 1990.

———. "Selective Arrests, an Ever-Expanding DNA Forensic Database, and the Specter of an Early-Twenty-First-Century Equivalent of Phrenology." In *DNA and the Criminal Justice System: The Technology of Justice*, edited by D. Lazer, 315–34. Cambridge, MA: MIT Press, 2004.

Dvoskin, J. A., and Heilbrun, K. "Risk Assessment and Release Decision-Making: Toward Resolving the Great Debate." *Journal of the American Academy of Psychiatry and the Law* 29 (2001): 6–10.

Dwyer, T. "One Juror between Terrorist and Death." *Washington Post*, May 12, 2006. Accessed June 24, 2020. www.washingtonpost.com/wp-dyn/content/article/2006/05/11/AR2006051101884.html.

Eaglin, J. M. "Against Neorehabilitation." *SMU Law Review* 66, no. 1 (2013): 189–226.

Electronic Privacy Information Center. Uniting and Strengthening America by Providing Appropriate Tools Required to Intercept and Obstruct Terrorism (USA PATRIOT ACT) Act of 2001, H.R. 3162, 107th Cong. (2001). Accessed June 24, 2020. https://epic.org/privacy/terrorism/hr3162.pdf.

Enders, W., and Sandler, T. "The Effectiveness of Antiterrorism Policies: A Vector-Autoregression-Intervention Analysis." *American Political Science Review* 87, no. 4 (1993): 829–44.

———. "What Do We Know about the Substitution Effect in Transnational Terrorism?" In *Research on Terrorism: Trends, Achievements and Failures*, edited by A. Silke, 119–37. London: Routledge, 2004.

Engels, F. *The Condition of the Working Class in England*. Oxford: Oxford University Press, 1999.

Epp, C. R., Maynard-Moody, S., and Haider-Markel, D. P. *Pulled Over: How Police Stops Define Race and Citizenship*. Chicago: Chicago University Press, 2014.

Ericson, R. V. "Ten Uncertainties of Risk-Management Approaches to Security." *Canadian Journal of Criminology and Criminal Justice* 48, no. 3 (2006): 345–57.

Eschner, Kat. "Lie Detectors Don't Work as Advertised and They Never Did." *Smithsonian Magazine*, February 2, 2017. www.smithsonianmag.com/smart-news/lie-detectors-dont-work-advertised-and-they-never-did-180961956/.

Eterno, J. A., and Silverman, E. B. "The New York City Police Department's Compstat: Dream or Nightmare?" *International Journal of Police Science and Management* 8, no. 3 (2006): 218–31.

Eterno, J. A., and Silverman, E. B. *The Crime Numbers Game: Management by Manipulation*. Boca Raton, FL: CRC Press, 2012.

Evans, E. P. *The Criminal Prosecution and Capital Punishment of Animals*. London: William Heinemann, 1906.

Evans, J. C. "Hijacking Civil Liberties: The USA Patriot Act of 2001." *Loyola University Chicago Law Journal* 33, no. 4 (2002): 933–90.

Ewing, C. P. "Preventive Detention and Execution: The Constitutionality of Punishing Future Crimes." *Law and Human Behavior* 15, no. 2 (1991): 139–63.

Fagan, J., and Davies, G. "Street Stops and Broken Windows: Terry, Race, and Disorder in New York City." *Fordham Law Review* 28, no. 2 (2000): 456–504.

Fagan, J., Geller, A., Davies, G., and West, V. "Street Stops and Broken Windows Revisited: The Demography and Logic of Proactive Policing in a Safe and Changing City." In *Race, Ethnicity, and Policing: New and Essential Readings*, edited by S. K. Rice and M. D. White, 309–48. New York: New York University Press, 2010.

Farrington, D. P., and Loeber, R. "Epidemiology of Juvenile Violence." *Child and Adolescent Psychiatric Clinics* 9, no. 4 (2000): 733–48.

Feeley, M. M., and Simon, J. "Actuarial Justice: The Emerging New Criminal Law." In *The Futures of Criminology*, edited by D. Nelkin, 173–201. London: Sage, 1994.

———. "The New Penology: Notes on the Emerging Strategy of Corrections and Its Implications." *Criminology* 30, no. 4 (1992): 449–74.

Feldman, G. "'Intelligence' and 'Evidence': Sovereign Authority and the Differences That Words Make." In *Bodies as Evidence: Security, Knowledge, and Power*, edited by M. Maguire, U. Rao, and N, Zurawski, 159–74. Durham, NC: Duke University Press, 2018.

Felson, M., and Eckert, M. A. *Crime and Everyday Life*. 6th ed. Los Angeles: Sage, 2018.

Ferrell, J. "Freight Train Graffiti: Subculture, Crime, Dislocation." *Justice Quarterly* 15, no. 4 (1998): 587–608.

———. "Urban Graffiti: Crime, Control, and Resistance." *Youth and Society* 27, no. 1 (1995): 73–92.

Ferrero, G. L. *Criminal Man, According to the Classification of Cesare Lombroso*. New York: G. P. Putnam, 1911.

Finkelhor, D., and Asdigian, N. L. "Risk Factors for Youth Victimization: Beyond a Lifestyles/Routine Activities Theory Approach." *Violence and Victims* 11, no. 1 (1996): 3–19.

Fisher, D. "Surveillance Auckland: 3140-Plus Cameras Are Watching Every Move—and Police Want More for the Suburbs. *New Zealand Herald*, November 25, 2015. Accessed February 11, 2019. www.nzherald.co.nz/nz/news/article.cfm?c_id=1&objectid=11550646.

Fisher, L. *In the Name of National Security: Unchecked Presidential Power and the Reynolds Case*. Lawrence: University Press of Kansas, 2006.

Floud, J. "Dangerousness and Criminal Justice." *British Journal of Criminology* 22 (1982): 213–28.

Foley, B. J. "Guantanamo and Beyond: Danger of Rigging the Rules." *Journal of Criminal Law and Criminology* 97, no. 4 (2007): 1009–69.

Foote, C. "Comments on Preventive Detention." *Journal of Legal Education* 23 (1970): 48–53.

Form, W. "Charging Waterboarding as War Crime: U.S. War Crime Trials in the Far East after World War II." *Chapman Journal of Criminal Justice* 2, no. 1 (2011): 247–73.

Forst, B. "Selective Incapacitation: A Sheep in Wolf's Clothing." *Judicature* 68, nos. 4–5 (1984): 153–60.

Foucault, M. *Discipline and Punish: The Birth of the Prison.* Translated by A. Sheridan. London: Penguin, 1977.

Foucault, M. *The Archaeology of Knowledge.* Translated by A. M. Sheridan Smith. New York: Pantheon Books, 1972.

Foucault, M. *Power/Knowledge: Selected Interviews and Other Writings, 1972–1977.* Edited by C. Gordon. New York: Pantheon Books, 1980.

Fox News. "McVeigh's Apr. 26 Letter to Fox News." April 26, 2001. Accessed June 24, 2020. www.foxnews.com/story/mcveighs-apr-26-letter-to-fox-news.

Frase, R. S., and Roberts, J. V. *Paying for the Past: The Case against Prior Record Sentence Enhancements.* New York: Oxford University Press, 2019.

Frederickson, H. G., and LaPorte, T. R. "Airport Security, High Reliability, and the Problems of Rationality." *Public Administration Review* 62 (2002): 34–44.

Friedman, L. M., and Percival, R. V. *The Roots of Justice: Crime and Punishment in Alameda County, California 1870–1910.* Chapel Hill: University of North Carolina Press, 2017.

Fussell, P. *Class: A Guide through the American Status System.* New York: Summit Books, 1983.

Garland, D. *The Culture of Control: Crime and Social Order in Contemporary Society.* Oxford: Oxford University Press, 2001.

Gascón, L. D., and Roussell, A. *The Limits of Community Policing: Civilian Power and Police Accountability in Black and Brown Los Angeles.* New York: New York University Press, 2019.

Geason. S. "Preventing Graffiti and Vandalism." Paper presented at Designing out Crime: Crime Prevention through Environmental Design (CPTED), Convened by the Australian Institute of Criminology and NRMA Insurance. Sydney, Australia, June 16, 1989.

Gerlach, N. *The Genetic Imaginary: DNA in the Canadian Criminal Justice System.* Toronto: University of Toronto Press, 2004.

Ghiglieri, M. *The Dark Side of Man: Tracing the Origins of Male Violence.* Reading, MA: Perseus Books, 1999.

Giannelli, P. C. "Daubert Challenges to Fingerprints." *Faculty Publications* 155 (2006): 624–42.

Gibbons, A. "Tracking the Evolutionary History of a 'Warrior' Gene." *Science* 304 (2004): 818.

Gibson, M. *Born to Crime: Cesare Lombroso and the Origins of Biological Criminology.* Westport, CT: Praeger, 2002.

Gill, M., and Spriggs, A. *Assessing the Impact of CCTV.* London: Home Office Research, Development and Statistics Directorate, 2005.

Gill, M., and Turbin, V. "CCTV and Shop Theft: Towards a Realistic Evaluation." In *Surveillance, Closed Circuit Television and Social Control,* edited by C. Norris, J. Moran, and G. Armstrong, 189–204. Aldershot, UK: Ashgate, 1998.

Global Commission on Drug Policy. *The War on Drugs.* Switzerland: Global Commission on Drug Policy, 2011. Accessed June 24, 2020. www.globalcommissionondrugs.org/reports/the-war-on-drugs.

Goold, B. J. "Public Area Surveillance and Police Work: The Impact of CCTV on Police Behavior and Autonomy." *Surveillance and Society* 1, no. 2 (2003): 191–203.

Gottfredson, D. M. *Exploring Criminal Justice: An Introduction*. Los Angeles: Roxbury Publishing Company, 1999.

Gottfredson, M., and Hirschi, T. "The True Value of Lambda Would Appear to Be Zero: An Essay on Career Criminals, Criminal Careers, Selective Incapacitation, Cohort Studies, and Related Topics." *Criminology* 24, no. 2 (1986): 213–34.

Gottfredson, S. D., and Gottfredson, D. M. "Selective Incapacitation?" *Annals of the American Academy of Political and Social Science* 478 (1985): 135–49.

Graham, K. "Overcharging." *Ohio State Journal of Criminal Law* 11 (2013): 701–24.

Gras, M. L. "The Legal Regulation of CCTV in Europe." *Surveillance and Society* 2, nos. 2/3 (2004): 216–29.

Greenberg, K. J., and Dratel, J. L. *The Torture Papers: The Road to Abu Ghraib*. New York: Cambridge University Press, 2005.

Greenwald, G. *No Place to Hide: Edward Snowden, the NSA, and the US Surveillance State*. New York: Metropolitan, 2014.

Greenwald, G. "Obama Killed a 16-Year-Old American in Yemen: Trump Just Killed His 8-Year-Old Sister." The Intercept, January 31, 2017. Accessed June 24, 2020. https://theintercept.com/2017/01/30/obama-killed-a-16-year-old-american-in-yemen-trump-just-killed-his-8-year-old-sister/.

Greenwood, P. W., and Abrahamse, A. F. *Selective Incapacitation*. Santa Monica, CA: Rand Corporation, 1982.

Grieve, D. "Possession of Truth." *Journal of Forensic Identification* 46 (1996): 521–28.

Guardian. "London Terror Attack: What We Know So Far." June 5, 2017. Accessed June 24, 2020. www.theguardian.com/uk-news/2017/jun/04/london-attacks-what-we-know-so-far-london-bridge-borough-market-vauxhall.

Haggerty, K. D. "Surveillance, Crime and the Police." In *Routledge Handbook of Surveillance Studies*, edited by K. Ball, K. Haggerty, and D. Lyon, 235–43. London: Routledge, 2012.

Hall, S., Winlow, S., and Ancrum, C. *Criminal Identities and Consumer Culture: Crime, Exclusion and the New Culture of Narcissism*. Cullompton, UK: Willan, 2008.

Hancox, P. D., and Morgan, J. B. "The Use of CCTV for Police Control at Football Matches." *Police Research Bulletin* 24 (1975): 41–44.

Hannah-Moffat, K. "Algorithmic Risk Governance: Big Data Analytics, Race and Information Activism in Criminal Justice Debates." *Theoretical Criminology* 23, no. 4 (2019): 453–70.

———. "Criminogenic Needs and the Transformative Risk Subject: Hybridizations of Risk/Need in Penality." *Punishment and Society* 7, no. 1 (2005): 29–51.

———. "Losing Ground: Gendered Knowledges, Parole Risk, and Responsibility." *Social Politics* 11, no. 3 (2004): 363–85.

Haraway, D. *Modest_Witness@Second_Millennium. FemaleMan_Meets_OncoMouse: Feminism and Technoscience*. New York: Routledge, 1997.

———. "Situated Knowledges: The Science Question in Feminism and the Privilege of Partial Perspective." *Feminist Studies* 14, no. 3 (1988): 575–99.

Harcourt, B. E. *Against Prediction: Profiling, Policing, and Punishing in an Actuarial Age*. Chicago: University of Chicago Press, 2007.

Harcourt, B. E., and Ludwig, J. "Reefer Madness: Broken Windows Policing and Misdemeanor Marijuana Arrests in New York." *John M. Olin Program in Law and Economics Working Paper* 317 (2006): 1–21.

Harding, S. "Introduction: Eurocentric Scientific Illiteracy—A Challenge for the World Community." In *The "Racial" Economy of Science: Toward a Democratic Future*, edited by S. Harding, 1–22. Bloomington: Indiana University Press, 1993.

———. *Objectivity and Diversity: Another Logic of Scientific Research*. Chicago: University of Chicago Press, 2015.

———. *The Science Question in Feminism*. Milton Keynes, UK: Open University Press, 1986.

Harvey, D. *A Brief History of Neoliberalism*. Oxford: Oxford University Press, 2005.

———. "The Fetish of Technology: Causes and Consequences." *Macalester International* 13, no. 7 (2003): 3–30.

Hattery, A. J., and Smith, E. *Policing Black Bodies: How Black Lives Are Surveilled and How to Work for Change*. Lanham, MD: Rowman and Littlefield, 2018.

Henne, K., and Shah, R. "Unveiling White Logic in Criminological Research: An Intertextual Analysis." *Contemporary Justice Review* 18, no. 2 (2015): 105–20.

Henne, K., and Troshynski, E. I. "Intersectional Criminologies for the Contemporary Moment: Crucial Questions of Power, Praxis and Technologies of Control." *Critical Criminology* 27, no. 1 (2019): 55–71.

Henry, S., and Milovanovic, D. *Constitutive Criminology: Beyond Postmodernism*. London: Sage, 1996.

Hirschi, T., and Gottfredson, M. "Age and the Explanation of Crime." *American Journal of Sociology* 89, no. 3 (1983): 552–84.

Hobbes, T. *Leviathan*. Edited by R. Tuck. New York: Cambridge University Press, 1996.

Holland, N., and DeLisi, M. "The Warrior Gene: MAOA Genotype and Antisocial Behavior in Males." In *The Routledge International Handbook of Biosocial Criminology*, edited by M. DeLisi and M. G. Vaughn, 179–89. New York: Routledge, 2015.

Home Office. "Official Statistics: National DNA Database Statistics." UK Government, 2019. Accessed January 14, 2020. www.gov.uk/government/statistics/national-dna-database-statistics.

Honigsberg, P. J. *A Place Outside the Law: Forgotten Voices from Guantánamo*. Boston: Beacon Press, 2019.

Hook, G. R. "'Warrior Genes' and the Disease of Being Māori." *MAI Review* 2 (2009): 1–11.

Horn, D. *The Criminal Body: Lombroso and the Anatomy of Deviance*. New York: Routledge, 2003.

"The Hound of the Baskervilles," *The Strand* 22, no. 128 (1901): 124.

Howell, K. B. "Broken Lives from Broken Windows: The Hidden Costs of Aggressive Order-Maintenance Policing." *New York University Review of Law and Social Change* 33 (2009): 271–329.

Hruska, R. L. "Preventive Detention: The Constitution and the Congress." *Creighton Law Review* 3 (1969): 36–87.

Hsu, H. Y., and McDowall, D. "Does Target-Hardening Result in Deadlier Terrorist Attacks against Protected Targets? An Examination of Unintended Harmful Consequences." *Journal of Research in Crime and Delinquency* 54, no. 6 (2017): 930–57.

Hulsman, L. H. C. 1986. "Critical Criminology and the Concept of Crime." *Contemporary Crises* 10, no. 1 (1986): 63–80.

Hunter, R. D., and Jeffery, C. R. "Preventing Convenience Store Robbery through Environmental Design." In *Situational Crime Prevention: Successful Case Studies*, edited by R. V. Clarke, 191–99. Monsey, NY: Criminal Justice Press, 1997.

Hussain, M. "Guantánamo Bay 'Forever' Prisoner Speaks Out—to Praise Congress, Lindsey Graham, and Thomas Friedman." The Intercept, March 18, 2019. Accessed June 24, 2020. https://theintercept.com/2019/03/17/guantanamo-bay-prisoners-al-sharbi/.

Hyatt, J. M., Bergstrom, M. H., and Chanenson, S. L. "Follow the Evidence: Integrate Risk Assessment into Sentencing." *Federal Sentencing Reporter* 23, no. 4 (2011): 266–68.

Indictment, United States v. Abdulmutallab. Case 2:10-cr-20005 (E.D. Mich.). January 6, 2010. Accessed June 24, 2020. www.cbsnews.com/htdocs/pdf/Abdulmutallab_Indictment.pdf.

Information Awareness Office. home page. n.d. https://web.archive.org/web/20020802012150/http://www.darpa.mil/iao/.

"Information Awareness Office Seal." *Wikipedia*. n.d. https://en.wikipedia.org/wiki/Information_Awareness_Office#/media/File:IAO-logo.png.

Jacobs, B. "The Manipulation of Fear in Carjacking." *Journal of Contemporary Ethnography* 42, no. 5 (2013): 523–44.

Jacobs, B. A., Topalli, V., and Wright, R. "Carjacking, Streetlife and Offender Motivation." *British Journal of Criminology* 43, no. 4 (2003): 673–88.

Jacobs, J. B. *The Eternal Criminal Record.* Cambridge, MA: Harvard University Press, 2015.

Jaffee, S. R., Caspi, A., Moffitt, T. E., Dodge, K. A., Rutter, M., Taylor, A., and Tully, L. A. "Nature x Nurture: Genetic Vulnerabilities Interact with Physical Maltreatment to Promote Conduct Problems." *Development and Psychopathology* 17, no. 1 (2005): 67–84.

Jaffer, J. "Balancing Power in the U.S. Response to External Threats: NSA Surveillance and Guantanamo Detention." *New York City Law Review* 10, no. 2 (2007): 361–67.

Jansen, B. "TSA Defends Full-Body Scanners at Airport Checkpoints." *USA Today*, March 2, 2016. Accessed August 27, 2017. www.usatoday.com/story/news/2016/03/02/tsa-defends-full-body-scanners-airport-checkpoints/81203030/.

Jeffery, C. R. *Crime Prevention through Environmental Design*. Beverly Hills, CA: Sage, 1971.

Jehl, D. "70 Die in Attack at Egypt Temple." *New York Times*, November 18, 1997. Accessed June 24, 2020. www.nytimes.com/1997/11/18/world/70-die-in-attack-at-egypt-temple.html.

Joh, E. E. "The Forgotten Threat: Private Policing and the State." *Indiana Journal of Global Legal Studies* 13, no. 2 (2006): 357–89.

Jordan, D. A. "Decrypting the Fourth Amendment: Warrantless NSA Surveillance and the Enhanced Expectation of Privacy Provided by Encrypted Voice Over Protocol." *Boston College Law Review* 47, no. 3 (2006): 505–46.

Junger-Tas, J., and Marshall, I. H. "The Self-Report Methodology in Crime Research." *Crime and Justice* 25 (1999): 291–367.

Kadidal, S. "NSA Surveillance: The Implications for Civil Liberties." *I/S: A Journal of Law and Policy for the Information Society* 10, no. 2 (2014): 433–79.

Kant, I. *Critique of Pure Reason*. London: Everyman, 1993.

Katyal, N. K., and Caplan, R. "The Surprisingly Stronger Case for the Legality of the NSA Surveillance Program: The FDR Precedent." *Stanford Law Review* 60, (2007): 1023–78.

Katz, J. *Seductions of Crime: Moral and Sensual Attractions in Doing Evil*. New York: Basic Books, 1988.

Kelling, G. L. "Crime Control, the Police, and Culture Wars: Broken Windows and Cultural Pluralism." *Perspectives on Crime and Justice* 2 (1997): 1–28.

Kelling, G. L., and Coles, C. M. *Fixing Broken Windows: Restoring and Reducing Crime in Our Communities*. New York: Martin Kessler Books, 1996.

Kierkegaard, S. *Fear and Trembling*. Translated by W. Lowrie. New York: Doubleday Anchor Books, 1954.

Kitrosser, H. "'Macro-Transparency' as Structural Directive: A Look at the NSA Surveillance Controversy." *Minnesota Law Review* 91 (2007): 1163–1208.

Kitsuse, J. I., and Cicourel. A. V. "A Note on the Uses of Official Statistics." *Social Problems* 11, no. 2 (1963): 131–39.

Kleiman, M., Ostrom, B. J., and Cheesman, F. L., II. "Using Risk Assessment to Inform Sentencing Decisions for Nonviolent Offenders in Virginia." *Crime and Delinquency* 53, no. 1 (2007): 106–32.

Klein, N. *The Shock Doctrine: The Rise of Disaster Capitalism*. New York: Metropolitan Books, 2007.

Knorr Cetina, K. D. *Epistemic Cultures: How the Sciences Make Knowledge*. Cambridge, MA: Harvard University Press, 1999.

Knutsson, J., and Kuhlhorn, E. "Macro-Measures against Crime: The Example of Check Forgeries." In *Situational Crime Prevention: Successful Case Studies*, edited by R. V. Clarke, 113–21. Monsey, NY: Criminal Justice Press, 1997.

Kovandzic, T., Sloan, J. J., III, and Vieraitis, L. M. "Unintended Consequences of Politically Popular Sentencing Policy: The Homicide Promoting Effects of 'Three Strikes' in U.S. Cities (1980–1999)." *Criminology and Public Policy* 1, no. 3 (2002): 399–424.

Kramer, R. "Differential Punishment of Similar Behavior: Sentencing Assault Cases in a Specialized Family Violence Court and 'Regular Sentencing' Courts." *British Journal of Criminology* 56, no. 4 (2016): 689–708.

———. "Moral Panics and Urban Growth Machines: Official Reactions to Graffiti in New York City, 1990–2005." *Qualitative Sociology* 33, no. 3 (2010): 297–311.

———. "Neoliberal States and 'Flexible Penality': Punitive Practices in District Courts." *New Zealand Sociology* 30, no. 2 (2015): 44–58.

Kramer, R., Rajah, V., and Sung, H. E. "Neoliberal Prisons and Cognitive Treatment: Calibrating the Subjectivity of Incarcerated Young Men to Economic Inequalities." *Theoretical Criminology* 17, no. 4 (2013): 535–56.

Kruse, C. *The Social Life of Forensic Evidence*. Oakland: University of California Press, 2015.

Kuhlhorn, E. "Housing Allowances in a Welfare Society: Reducing the Temptation to Cheat." In *Situational Crime Prevention: Successful Case Studies*, edited by R. V. Clarke, 235–41. Monsey, NY: Criminal Justice Press, 1997.

Kuhn, T. *The Structure of Scientific Revolutions*. Chicago: University of Chicago Press, 1970.

Landes, W. M. "An Economic Study of U.S. Aircraft Hijacking, 1961–1976." *Journal of Law and Economics* 21, no. 1 (1978): 1–31.

Landynski, J. W. *Search and Seizure and the Supreme Court: A Study in Constitutional Interpretation*. Baltimore, MD: Johns Hopkins University Press, 1966.

Latour, B. *Science in Action: How to Follow Scientists and Engineers through Society*. Cambridge, MA: Harvard University Press, 1987.

Lea, R., and Chambers, G. "Monoamine Oxidase, Addiction, and the 'Warrior' Gene Hypothesis." *New Zealand Medical Journal* 120 (2007): 1–6.

Leman-Langlois, S. "The Local Impact of Police Videosurveillance on the Social Construction of Security." In *Technocrime: Technology, Crime and Social Control*, edited by S. Leman-Langlois, 27–45. Milton Park, UK: Routledge, 2011.

Leo, R. A. "False Confessions: Causes, Consequences and Implications." *Journal of the American Academy of Psychiatry and the Law* 37 (2009): 332–43.

Lindesmith, A., and Levin, Y. "The Lombrosian Myth in Criminology." *American Journal of Sociology* 42, no. 5 (1937): 653–71.

Lippert, R., and Wilkinson, B. "Capturing Crime, Criminals and the Public's Imagination: Assembling *Crime Stoppers* and CCTV Surveillance." *Crime Media Culture* 6, no. 2 (2010): 131–52.

Lippert, R. K., and Walby, K. *A Criminology of Policing and Security Frontiers*. Bristol, UK: Bristol University Press, 2019.

Loader, I. "Thinking Normatively about Private Security." *Journal of Law and Society* 24, no. 3 (1997): 377–94.

Loftus, E. F., and Palmer, J. C. "Reconstruction of Automobile Destruction: An Example of the Interaction between Language and Memory." *Journal of Verbal Learning and Verbal Behavior* 13, no. 5 (1974): 585–89.

Logan, J. R., and Molotch, H. L. *Urban Fortunes: The Political Economy of Place*. Berkeley: University of California Press, 1987.

Lombroso, C. *Criminal Man*. Translated by N. Rafter and M. Gibson. Durham, NC: Duke University Press, 2006.

Luhmann, N. *Theories of Distinction: Redescribing the Descriptions of Modernity*. Stanford, CA: Stanford University Press, 2002.

L'uomo delinquente in rapporto all'antropologia, alla giurisprudenza ed alla psichiatria. Atlante/Cesare Lombroso, 5th ed. Torino: Fratelli Bocca, 1897.

Lynch, M., Cole, S. A., McNally, R., and Jordan, K. *Truth Machine: The Contentious History of DNA Fingerprinting*. Chicago: Chicago University Press, 2008.

Lyon, D. "Surveillance as Social Sorting: Computer Codes and Mobile Bodies." In *Surveillance as Social Sorting: Privacy, Risk and Digital Discrimination*, edited by D. Lyon, 13–30. London: Routledge, 2003.

Maas, P. *Serpico*. New York: Viking Press, 1973.

Magnet, S., and Rodgers, T. "Stripping for the State: Whole Body Imaging Technologies and the Surveillance of Othered Bodies." *Feminist Media Studies* 12, no. 1 (2012): 101–18.

Martin, L. L. "Bombs, Bodies, and Biopolitics: Securitizing the Subject at the Airport Security Checkpoint." *Social and Cultural Geography* 11, no. 1 (2010): 17–34.

Martinson, R. "What Works? Questions and Answers about Prison Reform." *The Public Interest* 35 (1974): 22–54.

Marvell, T. B., and Moody, C. E. "The Lethal Effects of Three Strikes Laws." *Journal of Legal Studies* 30, no. 1 (2001): 89–106.

Marx, K. *Capital: A Critique of Political Economy*. Chicago: Charles H. Kerr and Company, 1906.

———. "The Communist Manifesto." In *Karl Marx: Selected Writings*, edited by L. H. Simon, 157–86. Indianapolis, IN: Hackett, 1994.

———. "The German Ideology." In *Karl Marx: Selected Writings*, edited by L. H. Simon, 102–56. Indianapolis, IN: Hackett, 1994.

———. "Preface to *A Contribution to the Critique of Political Economy*." In *Karl Marx: Selected Writings*, edited by L. H. Simon, 209–13. Indianapolis, IN: Hackett, 1994.

Masco, J. P. "The Secrecy/Threat Matrix." In *Bodies as Evidence: Security, Knowledge, and Power*, edited by M. Maguire, U. Rao, and N. Zurawski, 175–200. Durham, NC: Duke University Press, 2018.

Masferrer, A. *Post 9/11 and the State of Permanent Legal Emergency: Security and Human Rights in Countering Terrorism*. New York: Springer, 2012.

Mason, C., and Magnet, S. "Surveillance Studies and Violence against Women." *Surveillance and Society* 10, no. 2 (2012): 105–18.

Massey, J. L., Krohn, M. D., and Bonati, L. M. "Property Crime and the Routine Activities of Individuals." *Journal of Research in Crime and Delinquency* 26, no. 4 (1989): 378–400.

Masuda, B. "Reduction of Employee Theft in a Retail Environment: Displacement vs. Diffusion of Benefits." In *Situational Crime Prevention: Successful Case Studies*, edited by R. V. Clarke, 183–90. Monsey, NY: Criminal Justice Press, 1997.

Matejik, L. A. "DNA Sampling: Privacy and Police Investigation in a Suspect Society." *Arkansas Law Review* 61, no. 1 (2009): 53–90.

Mathiesen, T. "Selective Incapacitation Revisited." *Law and Human Behavior* 22, no. 4 (1998): 455–69.

McIntosh, T., and Coster, S. "Indigenous Insider Knowledge and Prison Identity." *Counterfutures* 3 (2017): 69–98.

Mednick, S., and Christiansen, K. O. *Biosocial Bases of Criminal Behavior*. New York: Gardner, 1977.

Meehl, P. *Clinical versus Statistical Prediction: A Theoretical Analysis and a Review of the Evidence*. Minneapolis: University of Minnesota Press, 1954.

Merriman, T., and Cameron, V. "Risk-Taking: Behind the Warrior Gene Story." *New Zealand Medical Journal* 120 (2007): 1–4.

Merton, R. "Social Structure and Anomie." *American Sociological Review* 3, no. 5 (1938): 672–82.

Michaelson, R. "Roadside Bomb Injures Tourists on Bus Near Giza Pyramids." *The Guardian*, May 19, 2019. Accessed June 24, 2020. www.theguardian.com/world/2019/may/19/egypt-roadside-bomb-injures-tourists-near-giza-pyramids-officials-say.

Miethe, T. D., and Sousa, W. H. "Carjacking and Its Consequences: A Situational Analysis of Risk Factors for Differential Outcomes." *Security Journal* 23, no. 4 (2010): 241–58.

Miller, T. R., Cohen, M. A., and Wiersema, B. *Victim Costs and Consequences: A New Look* (NCJ 155282). U.S. Department of Justice, Office of Justice Programs, National Institute of Justice, 1996. Accessed June 24, 2020. www.ncjrs.gov/pdffiles/victcost.pdf.

Miller, W. *Cops and Bobbies: Police Authority in New York and London 1830–1870*. Chicago: University of Chicago Press, 1977.

———. "Lower-Class Culture as a Generating Milieu of Gang Delinquency." *Journal of Social Issues* 14, no. 3 (1958): 5–19.

Mills, C. W. *The Power Elite*. New York: Oxford University Press, 1956.

Mills, C. Wright. *The Sociological Imagination*. Harmondsworth, UK: Pelican Books, 1970.

Mnookin, J. L. "Fingerprint Evidence in an Age of DNA Profiling." *Brooklyn Law Review* 67, no. 1 (2001): 13–70.

Mocan, N., and Tekin, E. "Ugly Criminals." *Review of Economics and Statistics* 92, no. 1 (2010): 15–30.

Monahan, J. "A Jurisprudence of Risk Assessment: Forecasting Harm among Prisoners, Predators, and Patients." *Virginia Law Review* 92, no. 3 (2006): 391–435.

Monahan, J., and Skeem, J. L. "Risk Assessment in Criminal Sentencing." *Annual Review of Clinical Psychology* 12 (2016): 489–513.

Moore, M. H. "Sizing Up Compstat: An Important Administrative Innovation in Policing." *Criminology and Public Policy* 2, no. 3 (2003): 469–94.

Morris, H. "Persons and Punishment." *The Monist* 52 (1968): 475–501.

Morris, N. "'Dangerousness' and Incapacitation." In *A Reader on Punishment*, edited by A. Duff and D. Garland, 241–60. Oxford: Oxford University Press, 1994.

Morse, S. J. "Blame and Danger: An Essay on Preventive Detention." *Boston University Law Review* 76 (1996): 113–55.

Mrázek, R. *The Complete Lives of Camp People: Colonialism, Fascism, Concentrated Modernity*. Durham, NC: Duke University Press, 2020.

Muir, H., and Cowan, R. "Four Bombs in 50 Minutes: Britain Suffers Its Worst-Ever Terror Attack." *The Guardian*, July 8, 2005. Accessed June 24, 2020. https://web.archive.org/web/20071217222740/http://www.guardian.co.uk/uk_news/story/0%2C%2C1523819%2C00.html.

National Association of Criminal Defense Lawyers. *The Trial Penalty: The Sixth Amendment Right to Trial on the Verge of Extinction and How to Save It.* Washington, DC: National Association of Criminal Defense Lawyers, 2018. Accessed June 24, 2020. www.nacdl.org/trialpenaltyreport/.

National Commission on Terrorist Attacks upon the United States. *The 9/11 Commission Report.* New York: W. W. Norton, 2004.

Netter, B. "Using Group Statistics to Sentence Individual Criminals: An Ethical and Statistical Critique of the Virginia Risk Assessment Program." *Journal of Criminal Law and Criminology* 97, no. 3 (2007): 699–730.

Nietzsche, F. *The Will to Power.* Translated by W. Kaufmann and R. J. Hollingdale. New York: Vintage, 1968.

Norris, C. "The Success of Failure: Accounting for the Global Growth of CCTV." In *Routledge Handbook of Surveillance Studies*, edited by K. Ball, K. Haggerty, and D. Lyon, 251–58. London: Routledge, 2012.

Norris, C., McCahill, M., and Wood, D. "The Growth of CCTV: A Global Perspective on the International Diffusion of Video Surveillance in Publicly Accessible Space." *Surveillance and Society* 2, nos. 2/3 (2004): 110–35.

Oleson, J. C. "Blowing Out All the Candles: Some Thoughts on the Twenty-Fifth Anniversary of the Sentencing Reform Act of 1984." *University of Richmond Law Review* 45, no. 2 (2011): 693–763.

———. *Criminal Genius: A Portrait of High-IQ Offenders.* Oakland: University of California Press, 2016.

———. "'Drown the World': Imperfect Necessity and Total Cultural Revolution." *Unbound: Harvard Journal of the Legal Left* 3 (2007): 19–116.

———. "Habitual Criminal Legislation in New Zealand: Three Years of Three-Strikes." *Australian and New Zealand Journal of Criminology* 48, no. 2 (2011): 277–92.

———. "The New Eugenics: Black Hyper-Incarceration and Human Abatement." *Social Sciences* 5, no. 4, 66 (2016): 1–20.

———. "Risk in Sentencing: Constitutionally-Suspect Variables and Evidence-Based Sentencing." *SMU Law Review* 64 (2011): 1329–1404.

Oleson, J. C., VanBenschoten, S. W., Robinson, C. R., and Lowenkamp, C. T. "Training to See Risk: Measuring the Accuracy of Clinical and Actuarial Risk Assessments among Federal Probation Officers." *Federal Probation* 75, no. 2 (2011): 52–56.

O'Malley, P. *Crime and Risk.* Los Angeles: Sage, 2010.

———. "Risk, Power and Crime Prevention." *Economy and Society* 21, no. 3 (1992): 252–75.

Orwell, G. *Nineteen Eighty-Four.* London: Secker and Warburg, 1949.

Packer, H. *The Limits of the Criminal Sanction.* Stanford, CA: Stanford University Press, 1968.

Palmen, N. "Forget Carjacking, the Next Big Threat Is Car Hacking." *Automotive Industries*, August 2014, 38–40. Accessed February 15, 2019. www.ai-online.com/Adv/Previous/show_issue.php?id=6233#sthash.bGaPfi66.J4vgoUVU.dpbs.

Parks, L. "Points of Departure: The Culture of US Airport Screening." *Journal of Visual Culture* 6, no. 2 (2007): 183–200.

Paternoster, R., and Bachman, R. *Explaining Criminals and Crime: Essays in Contemporary Criminological Theory.* Los Angeles: Roxbury Publishing Company, 2001.

Pedersen, S. "China Trials 'Deadbeat Map' App to Monitor Citizens' Debts as Part of Social Credit Score System." *The Telegraph*, January 24, 2019. Accessed June 24, 2020. www.telegraph.co.uk/news/2019/01/24/china-trials-deadbeat-map-app-monitor-citizens-debts-part-social/.

Perry, S., Apel, R., Newman, G. R., and Clarke, R. V. "The Situational Prevention of Terrorism: An Evaluation of the Israeli West Bank Barrier." *Journal of Quantitative Criminology* 33, no. 4 (2017): 727–51.

Poole, R. W., Jr. "Airport Security: Time for a New Model." In *The Economic Costs and Consequences of Terrorism*, edited by H. W. Richardson, P. Gordon, and J. E. Moore II, 67–97. Cheltenham, UK: Edward Elgar, 2007.

Poyner, B. "Situational Crime Prevention in Two Parking Facilities." *Security Journal* 2, no. 2 (1991): 96–101.

Pratt, J., and Eriksson, A. *Contrasts in Punishment: An Explanation of Anglophone Excess and Nordic Exceptionalism.* London: Routledge, 2014.

Prenzler, T. *Police Corruption: Preventing Misconduct and Maintaining Integrity.* Boca Raton: CRC Press, 2009.

Quan, N. "Black and White or Red All Over? The Impropriety of Using Crime Scene DNA to Construct Racial Profiles of Suspects." *Southern California Law Review* 84 (2011): 1403–44.

Quinlan, A. "Technoscience and Affected Bodies." *International Journal of Gender, Science and Technology* 6, no. 3 (2015): 330–45.

Quinsey, V. L. "Evolutionary Theory and Criminal Behavior." *Legal and Criminological Psychology* 7, no. 1 (2002): 1–13.

Rafter, N. *Creating Born Criminals.* Chicago: University of Illinois Press, 1997.

———. *The Criminal Brain: Understanding Biological Theories of Crime.* New York: New York University Press, 2008.

———. "Criminology's Darkest Hour: Biocriminology in Nazi Germany." *Australian and New Zealand Journal of Criminology* 41, no. 2 (2008): 287–306.

Raine, A. *The Anatomy of Violence: The Biological Roots of Crime.* New York: Random House, 2013.

———. "Annotation: The Role of Prefrontal Deficits, Low Automatic Arousal, and Early Health Factors in the Development of Antisocial and Aggressive Behavior in Children." *Journal of Child Psychology and Psychiatry* 43, no. 4 (2002): 417–34.

Raine, A., Laufer, W. S., Yang, Y., Narr, K. L., Thompson, P., and Toga, A. W. "Increased Executive Functioning, Attention, and Cortical Thickness in White-Collar Criminals." *Human Brain Mapping* 33, no 12 (2012): 2932–40.

Ralph, J. G. *America's War on Terror: The State of the 9/11 Exception from Bush to Obama.* Oxford: Oxford University Press, 2013.

Rasdan, A. J. "The Moussaoui Case: The Mess from Minnesota." *William Mitchell Law Review* 31, no. 4 (2005): 1417–59.

Reiss, A. *The Police and the Public.* New Haven, CT: Yale University Press, 1971.

Reppetto, T. A. "Crime Prevention and the Displacement Phenomenon." *Crime and Delinquency* 22, no. 2 (1976): 166–77.

Rhee, S. H., and Waldman, I. D. "Genetic and Environmental Influences on Antisocial Behavior: A Meta-analysis of Twin and Adoption Studies." *Psychological Bulletin* 128, no. 3 (2002): 490–529.

Risen, J. *Pay Any Price: Greed, Power, and Endless War*. Boston: Mariner Books, 2015.

Robinson, P. H. "Punishing Dangerousness: Cloaking Preventive Detention as Criminal Justice." *Harvard Law Review* 114, no. 5 (2001): 1429–56.

Rocque, M., Welsh, B. C., and Raine, A. "Biosocial Criminology and Modern Crime Prevention." *Journal of Criminal Justice* 40 (2012): 306–12.

Roncek, D. W., and Maier, P. A. "Bars, Blocks, and Crimes Revisited: Linking the Theory of Routine Activities to the Empiricism of 'Hot Spots.'" *Criminology* 29, no. 4 (1991): 725–53.

Rosen, C. "Liberty, Privacy, and DNA Databases." *New Atlantis: A Journal of Technology and Society* (Spring 2003): 37–52.

Rosenberg, C. "Trial for Men Accused of Plotting 9/11 Attacks Is Set for 2021." *New York Times*, August 30, 2019. Accessed June 24, 2020. www.nytimes.com/2019/08/30/us/politics/sept-11-trial-guantanamo-bay.html.

Roussell, A. "Policing the Anticommunity: Race, Deterritorialization, and Labor Market Reorganization in South Los Angeles." *Law and Society Review* 49, no. 4 (2015): 813–45.

Roussell, A., and Gascón, L. D. "Defining 'Policeability': Cooperation, Control, and Resistance in South Los Angeles Community-Police Meetings." *Social Problems* 61, no. 2 (2014): 237–58.

Rowe, D. "An Adaptive Strategy Theory of Crime and Delinquency." In *Delinquency and Crime: Current Theories*, edited by J. D. Hawkins, 268–314. Cambridge: Cambridge University Press, 1996.

———. *Biology and Crime*. Los Angeles, CA: Roxbury, 2002.

Rowe, D. C. "Biometrical Genetic Models of Self-Reported Delinquent Behavior: A Twin Study." *Behavior Genetics* 13, no. 5 (1983): 473–89.

Russell, S. "The Failure of Postmodern Criminology." *Critical Criminology* 8, no. 2 (1997): 61–90.

Samenow, S. E. *Inside the Criminal Mind*. New York: Random House, 1984.

———. *The Myth of the Out of Character Crime*. Westport, CT: Praeger, 2007.

Sampsell-Jones, T. "Preventive Detention, Character Evidence, and the New Criminal Law." *Utah Law Review* 3 (2010): 723–77.

Samti, F., and Gall, C. "Scores Die in Attack at Tunisian Beach Hotel." *New York Times*, June 27, 2015, A9.

Sarre, R. "Researching Private Policing: Challenges and Agendas for Researchers." *Security Journal* 18, no. 3 (2005): 57–70.

Schmitt, C. *Political Theology: Four Chapters on the Concept of Sovereignty*. Translated by G. Schwab. Chicago: University of Chicago Press, 1985.

Schneier, B. *Beyond Fear: Thinking Sensibly about Security in an Uncertain World*. New York: Copernicus Books, 2003.

Schwaninger, A. "Increasing Efficiency in Airport Security Screening." *WIT Transactions on the Built Environment* 82 (2005): 405–16.

Schweitzer, N. J., and Saks, M. J. "The *CSI* Effect: Popular Fiction about Forensic Science Affects the Public's Expectations about Real Forensic Science." *Jurimetrics* 47, no. 3 (2007): 357–64.

Seidenstat, P. "Terrorism, Airport Security, and the Private Sector." *Review of Policy Research* 21, no. 3 (2004): 275–91.

Select Committee on Home Affairs. "Nature and Extent of Young Black People's Overrepresentation." Home Affairs—Second Report. UK Parliament, 2007. Accessed January 14, 2020. https://publications.parliament.uk/pa/cm200607/cmselect/cmhaff/181/18105.htm.

Sellin, T., and Wolfgang, M. E. *The Measurement of Delinquency*. New York: Wiley, 1964.

Shane, S. "U.S. Approves Targeted Killing of American Cleric." *New York Times*, April 6, 2010, A12. Accessed June 24, 2020. https://web.archive.org/web/20100408031248/http://www.nytimes.com//2010//04//07//world//middleeast//07yemen.html.

Shannon, L. W. "Risk Assessment vs Real Prediction: The Prediction Problem and Public Trust." *Journal of Quantitative Criminology* 1, no. 2 (1985): 159–89.

Shearing, C., and Stenning, P. "Modern Private Security: Its Growth and Implications." *Crime and Justice* 3 (1981): 193–245.

Sherman, L. W., and Weisburd, D. "General Deterrent Effects of Police Patrol in Crime 'Hot Spots': A Randomized, Controlled Trial." *Justice Quarterly* 12, no. 4 (1995): 625–48.

Shermer, L., and Johnson, B. "Criminal Prosecutions: Examining Prosecutorial and Charge Reductions in US Federal District Courts." *Justice Quarterly* 27, no. 3 (2010): 394–430.

Shichor, D. "Three Strikes as a Public Policy: The Convergence of the New Penology and the McDonaldization of Punishment." *Crime and Delinquency* 43, no. 4 (1997): 470–92.

"Shoe Bomb Suspect to Remain in Custody." CNN, December 25, 2001. Accessed August 27, 2017. http://edition.cnn.com/2001/US/12/24/investigation.plane/index.html.

Silver, E., and Miller, L. L. "A Cautionary Note on the Use of Actuarial Risk Assessment Tools for Social Control." *Crime and Delinquency* 48, no. 1 (2002): 138–61.

Silverman, E. B. *NYPD Battles Crime: Innovative Strategies in Policing*. Boston: Northeastern University Press, 1999.

Simon, J. "Mass Incarceration: From Social Policy to Social Problem." In *The Oxford Handbook of Sentencing and Corrections*, edited by J. Petersilia and K. Reitz, 23–52. New York: Oxford University Press, 2012.

Simoncelli, T. "Dangerous Excursions: The Case against Expanding Forensic DNA Databases to Innocent Persons." *Journal of Law, Medicine and Ethics* 34, no. 2 (2006): 390–97.

Sivarajasingam, V., Shepherd, J. P., and Matthews, K. "Effect of Urban Closed Circuit Television on Assault Injury and Violence Detection." *Injury Prevention* 9 (2003): 312–16.

Skeem, J. "Risk Technology in Sentencing: Testing the Promises and Perils (Commentary on Hannah-Moffat, 2011)." *Justice Quarterly* 30, no. 2 (2013): 297–303.

Sklansky, D. A. "Is the Exclusionary Rule Obsolete?" *Ohio State Journal of Criminal Law* 5 (2008): 567–84.

Skolnick, J. H., and Fyfe, J. J. *Above the Law: Police and the Excessive Use of Force*. New York: Free Press, 1993.

Slobogin, C. "Testilying: Police Perjury and What to Do about It." *University of Colorado Law Review* 67 (1996): 1037–60.

Smith, A. *An Inquiry into the Nature and Causes of the Wealth of Nations*. Edited by K. Sutherland. New York: Oxford University Press, 2008.

Smith, G. J. D., and O'Malley, P. "Driving Politics: Data-Driven Governance and Resistance." *British Journal of Criminology* 57, no. 2 (2017): 275–98.

Smith, N. *The New Urban Frontier: Gentrification and the Revanchist City*. London: Routledge, 1996.

Solzhenitsyn, A. *Cancer Ward*. Translated by N. Bethell and D. Burg. London: Vintage, 2003.

Somaiya, R. "Tracking the News on Air Cargo Explosives" *New York Times*, October 30, 2010. Accessed June 24, 2020. https://web.archive.org/web/20101104080318/http://thelede.blogs.nytimes.com/2010/10/30/tracking-the-news-on-air-cargo-explosives/.

Spencer, D. "Sex Offender as Homo Sacer." *Punishment and Society* 11, no. 2 (2009): 219–40.

Spohn, C. "The Effects of the Offender's Race, Ethnicity, and Sex on Federal Sentencing Outcomes in the Guidelines Era." *Law and Contemporary Problems* 76 (2013): 75–104.

Stanko, E. "Warnings to Women: Police Advice and Women's Safety in Britain." *Violence Against Women* 2, no. 1 (1996): 5–24.

Statista. *Revenue of G4S Worldwide from 2010 to 2017 (in Billion GBP)*. Released April 2018. Accessed February 11, 2019. www.statista.com/statistics/323193/revenue-of-g4s-worldwide/.

Steffensmeier, D., Ulmer, J., and Kramer, J. "The Interaction of Race, Gender, and Age in Criminal Sentencing: The Punishment Cost of Being Young, Black, and Male." *Criminology* 36, no. 4 (1998): 763–98.

Steyn, J. "Guantánamo Bay: The Legal Black Hole." *International and Comparative Law Quarterly* 53, no. 1 (2004): 1–15.

Stuntz, W. *The Collapse of American Criminal Justice*. Cambridge, MA: Harvard University Press, 2011.

Surette, R. *Media, Crime, and Criminal Justice: Images, Realities, and Policies*. 5th ed. Stamford, CT: Cengage Learning, 2015.

Taibbi, M. *The Divide: American Injustice in the Age of the Wealth Gap*. New York: Spiegel and Grau, 2014.

Taifa, N. "Three-Strikes-and-You're-Out: Mandatory Life Imprisonment for Third Time Felons." *University of Dayton Law Review* 20, no. 2 (1995): 717–25.

Taub, B. "Guantánamo's Darkest Secret." *New Yorker*, April 15, 2019. Accessed June 24, 2020. www.newyorker.com/magazine/2019/04/22/guantanamos-darkest-secret.

Taylor, N. "State Surveillance and the Right to Privacy." *Surveillance and Society* 1, no. 1 (2002): 66–85.

"Terrorist Hid Explosives in His Bottom." *The Telegraph*, September 21, 2009. Accessed August 28, 2017. www.telegraph.co.uk/news/newstopics/howaboutthat/6212908/Terrorist-hid-explosives-in-his-bottom.html.

Thornhill, R., and Palmer, C. T. *A Natural History of Rape: Biological Bases of Sexual Coercion*. Cambridge, MA: MIT Press, 2000.

Tilley, N. "Evaluating the Effectiveness of CCTV Schemes." In *Surveillance, Closed Circuit Television and Social Control*, edited by C. Norris, J. Moran and G. Armstrong, 139–54. Aldershot, UK: Ashgate, 1998.

Toffler, A. *Future Shock*. New York: Random House, 1970.

Tombs, S., and Whyte, D. *The Corporate Criminal: Why Corporations Must Be Abolished*. New York: Routledge, 2015.

Tonry, M. "Ethnicity, Crime, and Immigration." *Crime and Justice* 21 (1997): 1–29.

Tonry, M., and Lynch, M. "Intermediate Sanctions." *Crime and Justice* 20 (1996): 99–144.

Tseloni, A., Wittebrood, K., Farrell, G., and Pease, K. "Burglary Victimization in England and Wales, the United States and the Netherlands." *British Journal of Criminology* 44, no. 1 (2004): 66–91.

Tuvblad, C., Grann, M., and Lichtenstein, P. "Heritability for Adolescent Antisocial Behavior Differs with Socioeconomic Status: Gene-Environment Interaction." *Journal of Child Psychology and Psychiatry* 47, no. 7 (2006): 734–43.

Tyler, T. R. *Why People Obey the Law*. New Haven, CT: Yale University Press, 1990.

Van Cleve, N. G., and Mayes, L. "Criminal Justice through 'Colorblind' Lenses: A call to Examine the Mutual Constitution of Race and Criminal Justice." *Law and Social Inquiry* 40, no. 2 (2015): 406–32.

van Eijk, G. "Socioeconomic Marginality in Sentencing: The Built-In Bias in Risk Assessment Tools and the Reproduction of Social Inequality." *Punishment and Society* 19, no. 4 (2017): 463–81.

van Steden, R., and Sarre, R. "The Growth of Private Security: Trends in the European Union." *Security Journal* 20 (2007): 222–35.

von Hirsch, A. "Prediction of Criminal Conduct and Preventive Confinement of Convicted Persons." *Buffalo Law Review* 21, (1972): 717–58.

von Hirsch, A. "Selective Incapacitation Re-examined: The National Academy of Sciences' Report on Criminal Careers and 'Career Criminals.'" *Criminal Justice Ethics* 7 (1988): 19–34.

von Hirsch, A., and Gottfredson, D. M. "Selective Incapacitation: Some Queries about Research Design and Equity." *New York University Review of Law and Social Change* 12, no. 1 (1984): 11–51.

Wacquant, L. "Crafting the Neoliberal State: Workfare, Prisonfare, and Social Insecurity." *Sociological Forum* 25, no. 2 (2010): 197–220.

———. *Prisons of Poverty*. Minneapolis: University of Minnesota Press, 2009.

Wagner, P., and Rabuy, B. "Following the Money of Mass Incarceration." Prison Policy Initiative, January 25, 2017. Accessed June 24, 2020. www.prisonpolicy.org/reports/money.html.

Wakefield, A. "The Public Surveillance Functions of Private Security." *Surveillance and Society* 2, no. 4 (2005): 529–45.

Walby, K., and Carrier, N. "The Rise of Biocriminology: Capturing Observable Bodily Economies of 'Criminal Man.'" *Criminology and Criminal Justice* 10, no. 3 (2010): 261–85.

Walker, S. "Broken Windows and Fractured History: The Use and Misuse of History in Recent Police Patrol Analysis." *Justice Quarterly* 75 (1985): 75–90.

Walklate, S. "Risk and Criminal Victimization: A Modernist Dilemma?" *British Journal of Criminology* 37, no. 1 (1997): 35–45.

Wallach, E. "Drop by Drop: Forgetting the History of Water Torture in US Courts." *Columbia Journal of Transnational Law* 45 (2007): 468–506.

Walsh, A. "Crazy by Design: A Biosocial Approach to the Age-Crime Curve." In *Biosocial Criminology: New Directions in Theory and Research*, edited by A. Walsh and K. M. Beaver, 154–75. New York: Routledge, 2009.

———. "Evolutionary Psychology and Criminal Behavior." In *Missing the Revolution: Darwinism for Social Scientists*, edited by J. H. Barkow, 225–68. Oxford: Oxford University Press, 2006.

———. *Race and Crime: A Biosocial Analysis*. New York: Nova Science Publishers, 2004.

Walsh, A., and Beaver, K. M. "Introduction to Biosocial Criminology." In *Biosocial Criminology: New Directions in Theory and Research*, edited by A. Walsh and K. M. Beaver, 7–28. New York: Routledge, 2009.

Walsh, A., and Wright, J. P. "Rage against Reason: Addressing Critical Critics of Biosocial Research." *Journal of Theoretical and Philosophical Criminology* 7, no. 2 (2015): 61–72.

Walsh, A., and Yun, I. "Race and Criminology in the Age of Genomic Science." *Social Science Quarterly* 92, no. 5 (2011): 1279–96.

Walsh, W. F. "Compstat: An Analysis of an Emerging Police Managerial Paradigm." *Policing: An International Journal Police Strategies and Management* 24, no. 3 (2001): 347–62.

Walters. R. "Bhopal, Corporate Crime and Harms of the Powerful." *Global Social Policy* 9, no. 3 (2009): 324–27.

Weisburd, D., Mastrofski, S. D., Greenspan, R., and Willis, J. J. "The Growth of Compstat in American Policing." *Police Foundation Reports* (April 2004): 1–17.

Weisburd, D., Mastrofski, S. D., McNally, A. M., Greenspan, R., and Willis, J. J. "Reforming to Preserve: Compstat and Strategic Problem Solving in American Policing." *Criminology and Public Policy* 2, no. 3 (2003): 421–56.

Werth, R. "Individualizing Risk: Moral Judgement, Professional Knowledge and Affect in Parole Evaluations." *British Journal of Criminology* 57, no. 4 (2017): 808–27.

———. "Risk and Punishment: The Recent History and Uncertain Future of Actuarial, Algorithmic, and 'Evidence-Based' Penal Techniques." *Sociology Compass* 13, no. 2 (2019): 1–19.

———. "Theorizing the Performative Effects of Penal Risk Technologies: (Re) producing the Subject Who Must Be Dangerous." *Social and Legal Studies* 28, no. 3 (2019): 327–48.

Western, B. *Punishment and Inequality in America*. New York: Russell Sage Foundation, 2006.

White, R. "Graffiti, Crime Prevention and Cultural Space." *Current Issues in Criminal Justice* 12, no. 3 (2001): 253–68.

Whitehead, J. A., and Aden, S. H. "Forfeiting 'Enduring Freedom' for 'Homeland Security': A Constitutional Analysis of the USA Patriot Act and the Justice Department's Anti-terrorism Initiatives." *American University Law Review* 51, no. 6 (2002): 1081–1133.

Wikileaks. "Vault 7: CIA Hacking Tools Revealed." March 7, 2017. Accessed February 11, 2019. https://wikileaks.org/ciav7p1/.

Wikimedia commons. Fire Triangle. n.d. https://upload.wikimedia.org/wikipedia/commons/thumb/6/63/Fire_triangle_2.svg/3614px-Fire_triangle_2.svg.png.

Wikipedia. S.v. "G4S." Accessed February 11, 2019. https://en.wikipedia.org/wiki/G4S.

Wilkins, L. T. "The Politics of Prediction." In *Prediction in Criminology*, edited by D. P. Farrington and R. Tarling, 34–51. Albany: State University of New York Press, 1985.

Wilkinson, R., and Pickett, K. *The Spirit Level: Why Equality Is Better for Everyone*. 2nd ed. New York: Penguin, 2010.

Willis, J. J., Mastrofski, S. D., and Weisburd, D. "Compstat and Bureaucracy: A Case Study of Challenges and Opportunities for Change." *Justice Quarterly* 21, no. 3 (2004): 463–96.

———. "Making Sense of COMPSTAT: A Theory-Based Analysis of Organizational Change in Three Police Departments." *Law and Society Review* 41, no. 1 (2007): 147–88.

Wilson, D., and McCulloch, J. *Pre-crime: Pre-emption, Precaution and the Future*. London: Routledge, 2015.

Wilson, D., and Sutton, A. "Watched Over or Over-Watched? Open Street CCTV in Australia." *Australian and New Zealand Journal of Criminology* 37, no. 2 (2004): 211–30.

Wilson, J. Q., and Herrnstein, R. *Crime and Human Nature*. New York: Simon and Schuster, 1985.

Wilson, J. Q., and Kelling, G. L. "Broken Windows: The Police and Neighborhood Safety." *Atlantic Monthly* 249 (1982): 1–9. Accessed August 3, 2009. www.theatlantic.com/doc/198203/broken-windows.

Wing, F. G. "Putting the brakes on carjacking or accelerating it? The anti car theft act of 1992." *University of Richmond Law Review* 28, (1994): 385–441.

Withnall, A., and Lichfield, J. "Charlie Hebdo Shooting: At Least 12 Killed as Shots Fired at Satirical Magazine's Paris Office." *Independent*, January 7, 2015. Accessed June 24, 2020. www.independent.co.uk/news/world/europe/charlie-hebdo-shooting-10-killed-as-shots-fired-at-satirical-magazine-headquarters-according-to-reports-9962337.html.

Wolfgang, M., Figlio, R., and Sellin, T. *Delinquency in a Birth Cohort*. Chicago: Chicago University Press, 1972.

Wray, C. *Preliminary Semiannual Uniform Crime Report, January–June, 2019*. Washington, DC: United States Department of Justice, 2019. Accessed June 24, 2020. https://ucr.fbi.gov/crime-in-the-u.s/2019/preliminary-report.

Wright, J. P. "Inconvenient Truths: Science, Race, and Crime." In *Biosocial Criminology: New Directions in Theory and Research*, edited by A. Walsh and K. M. Beaver, 137–53. New York: Routledge, 2009.

Yang, M., Wong, S. C. P., and Coid, J. "The Efficacy of Violence Prediction: A Meta-analytic Comparison of Nine Risk Assessment Tools." *Psychological Bulletin* 136, no. 5 (2010): 740–67.

Zabell, S. L. "Fingerprint Evidence." *Journal of Law and Policy* 13, no. 1 (2005): 143–79.

Zimring, F. "Populism, Democratic Government, and the Decline of Expert Authority: Some Reflections on Three Strikes in California." *Pacific Law Journal* 28 (1996): 243–56.

Zimring, F., and Hawkins, G. *The Scale of Imprisonment*. Chicago: University of Chicago Press, 1991.

Index

Founded in 1893,
UNIVERSITY OF CALIFORNIA PRESS
publishes bold, progressive books and journals on topics in the arts, humanities, social sciences, and natural sciences—with a focus on social justice issues—that inspire thought and action among readers worldwide.

The UC PRESS FOUNDATION
raises funds to uphold the press's vital role as an independent, nonprofit publisher, and receives philanthropic support from a wide range of individuals and institutions—and from committed readers like you. To learn more, visit ucpress.edu/supportus.

Lightning Source UK Ltd.
Milton Keynes UK
UKHW010744220522
403312UK00002B/56